Digital Circuit Testing
and Testability

Digital Circuit Testing
and Testability

Parag K. Lala
Electrical Engineering Department
North Carolina Agricultural and Technical State University
Greensboro, North Carolina

ACADEMIC PRESS

San Diego   London   Boston
New York   Sydney   Tokyo   Toronto

Copyright © 1997 by Academic Press

ACADEMIC PRESS
525 B Street, Suite 1900, San Diego, CA 92101-4495, USA
1300 Boylston Street, Chestnut Hill, MA 02167, USA
http://www.apnet.com

ACADEMIC PRESS LIMITED
24–28 Oval Road, London NW1 7DX, UK
http://www.hbuk.co.uk/ap/

TK
7874.75
L35
1997

Library of Congress Cataloging-in-Publication Data

Lala, Parag K.
    Digital circuit testing and testability / Parag K. Lala.
        p.   cm.
    Includes bibliographical references and index.
    ISBN 0-12-434330-9 (alk. paper)
    1. Integrated circuits—Very large scale integration—
Testing.   2. Digital integrated circuits—Testing.
3. Integrated circuits—Fault tolerance.   I. Title.
TK7874.75.L35   1997
621.39'5'0287—dc20   96-42114
                                                         CIP

Printed in the United States of America
96 97 98 99 00 BB 9 8 7 6 5 4 3 2 1

To Meena

"What might have been is an abstraction
Remaining a perpetual possibility
Only in a world of speculation.
What might have been and what has been
Points to one end, which is always present"

T. S. Eliot (*Burnt Norton*, 1935)

Contents

Preface

Advances in VLSI (Very Large Scale Integration) technology have enabled the implementation of complex digital circuits/systems in single chips, reducing system size and power consumption. This in turn has intensified the complexity of testing such chips to verify that they function correctly. In general, special design techniques have to be used to make a chip fully testable. Research over the years has produced efficient techniques for both test generation and design for testability. This book provides a detailed coverage of many of these techniques; the emphasis is more on the concept rather than the nuts and bolts of each technique. It is primarily intended for those involved in VLSI chip design and testing. The material covered in the book should also be sufficient for senior-level undergraduate and first-year graduate students in Electrical Engineering and Computer Science who are taking a course in digital circuit testing. It is assumed that the readers of the book have sufficient background in switching theory and logic design.

This book has seven chapters. Chapter 1 deals with various types of faults that may occur in VLSI-based digital circuits. The modeling of faults at the gate level and at the transistor level are considered.

Chapter 2 examines in detail various techniques available for efficient fault detection in combinational logic circuits. The Boolean difference technique, although rarely used in practice, is covered for pedagogical reasons.

Chapter 3 covers techniques that can be used to enhance testability of combinational circuits. In recent years, the main thrust of research in this area has been on design techniques that guarantee testable circuits. Several such techniques are discussed in this chapter. Also, techniques for implementing testable PLA (Programmable Logic Array) design are covered.

Chapter 4 deals with various techniques available for test generation of sequential circuits. The testing of such circuits remains a major problem, for which no generally accepted solution has been found. The state table verification approach is covered for pedagogical reasons. Several recent circuit-based test generation techniques are discussed. Also, techniques based on functional fault models are thoroughly covered.

Chapter 5 focuses on the various design techniques that can be used to make sequential circuits easily testable. Testing of large sequential circuits, without some design modifications for improving testability, is for all practical purposes an almost impossible task. This chapter introduces first the key concepts of testability, followed by some ad hoc design-for-testability rules. Major design methods for incorporating testability in sequential circuits used in VLSI-based digital systems are discussed in detail.

Chapter 6 deals with test generation and response evaluation techniques used in BIST (Built-In Self-Test) schemes for VLSI chips. Because LFSR (Linear Feedback Shift Register)–based techniques are used in practice to generate test patterns and evaluate output responses in BIST, such techniques are thoroughly discussed. Also, some popular BIST architectures are examined.

Chapter 7 starts with a discussion of fault modeling for RAM (Random Access Memory) memory chips. Test generation schemes for an important class of faults in memory are discussed. Also, detailed coverage of the techniques available for enhancing the testability of RAM chips, and embedded RAM blocks in VLSI chips, is included.

I would like to thank my colleague Dr. Fadi Busaba for his help in preparing certain sections of the book. In addition, I am greatly indebted to my wife Meena and children Nupur and Kunal for their patience, tolerance, and understanding during the time it took to write the book.

The author can be contacted by e-mail (lala@ncat.edu).

Digital Circuit Testing
and Testability

Chapter 1 | Faults in Digital Circuits

1.1 Failures and Faults

A *failure* is said to have occurred in a circuit or system if it deviates from its specified behavior [1.1]. A *fault*, on the other hand, is a physical defect that may or may not cause a failure. A fault is characterized by its *nature, value, extent, and duration* [1.2]. The nature of a fault can be classified as logical or nonlogical. A *logical* fault causes the logic value at a point in a circuit to become opposite to the specified value. *Nonlogical* faults include the rest of the faults, such as the malfunction of the clock signal, power failure, and so forth. The value of a logical fault at a point in the circuit indicates whether the fault creates fixed or varying erroneous logical values. The extent of a fault specifies whether the effect of the fault is localized or distributed. A local fault affects only a single variable, whereas a distributed fault affects more than one. A logical fault, for example, is a local fault, whereas the malfunction of the clock is a distributed fault. The duration of a fault refers to whether the fault is permanent or temporary.

1.2 Modeling of Faults

Faults in a circuit may occur due to defective components, breaks in signal lines, lines shortened to ground or power supply, short-circuiting of signal lines, excessive delays, and so forth. Besides errors or ambiguities in design specifications, design rule violations, among other things, also result in faults. Faulkner *et al.* [1.3] have found that specification faults and design rule violations accounted for

1

10% of the total faults encountered during the commissioning of subsystems of a midrange mainframe computer implemented using MSI (Medium Scale Integration); however, during the system validation such faults constituted 44% of the total. Poor designs may also result in hazards, races, or metastable flip-flop behavior in a circuit; such faults manifest themselves as "intermittents" throughout the life of the circuit.

In general, the effect of a fault is represented by means of a model, which represents the change the fault produces in circuit signals. The fault models in use today are

1. Stuck-at fault
2. Bridging fault
3. Stuck-open fault

1.2.1 STUCK-AT FAULTS

The most common model used for logical faults is the *single stuck-at fault*. It assumes that a fault in a logic gate results in one of its inputs or the output being fixed to either a logic 0 (stuck-at-0) or a logic 1 (stuck-at-1). Stuck-at-0 and stuck-at-1 faults are often abbreviated to s-a-0 and s-a-1, respectively, and these abbreviations will be adopted here.

Let us consider a NAND gate with input A s-a-1 (Fig. 1.1). The NAND gate perceives the input A as a 1 irrespective of the logic value placed on the input. The output of the NAND gate in Fig. 1.1 is 0 for the input pattern shown, when the s-a-1 fault is present. The fault-free gate has an output of 1. Therefore, the pattern shown in Fig. 1.1 can be used as a *test* for the A input s-a-1, because there is a difference between the output of the fault-free and the faulty gate.

The stuck-at model, often referred to as *classical* fault model, offers good representation for the most common types of failures, for example, short-circuits (*shorts*) and open circuits (*opens*) in many technologies. Fig. 1.2 illustrates the CMOS (Complementary Metal Oxide Semiconductor) realization of a NAND gate, the numbers 1, 2, 3, and 4 indicating places where opens have occurred.

Figure 1.1 NAND gate with input A s-a-1

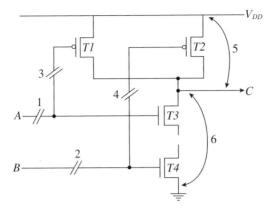

Figure 1.2   CMOS realization of two-input NAND gate

The numbers 5 and 6 identify the short between the output node and the ground, and the short between the output node and the V_{DD}, respectively. A short in a CMOS results if not enough metal is removed by the photolithography, whereas overremoval of metal results in an open circuit [1.4].

Fault 1 in Fig. 1.2 will disconnect input A from the gate of transistors $T1$ and $T3$. It has been shown that in such a situation one transistor may conduct and the other remain nonconducting [1.5]. Thus, the fault can be represented by a stuck-at value of A; if A is s-a-0, $T1$ will be ON and $T3$ OFF, and if A is s-a-1, 1, $T1$ will be OFF and $T3$ ON. In the presence of fault 3, only transistor $T3$ will remain ON, and hence the fault cannot be represented by the stuck-at model (see the paragraph stuck-on faults in Sec. 1.2.3). Faults 2 and 4 behave similarly as faults 1 and 3, respectively.

Fault 5 forces the output node to be shorted to V_{DD}, that is, the fault can be considered as s-a-1 fault. Similarly, fault 6 forces the output node to be s-a-0.

The stuck-at model is also used to represent multiple faults in circuits. In a *multiple stuck-at fault,* it is assumed that more than one signal line in the circuit are stuck at logic 1 or logic 0; in other words, a group of stuck-at faults exist in the circuit at the same time. A variation of the multiple fault is the *unidirectional fault.* A multiple fault is unidirectional if all of its constituent faults are either s-a-0 or s-a-1 but not both simultaneously. The stuck-at model has gained wide acceptance in the past mainly because of its relative success with small scale integration. However, it is not very effective in accounting for all faults in present day VLSI (Very Large Scale Integration), which mainly uses CMOS technology. Faults in CMOS circuits do not necessarily produce logical faults that can be described as stuck at faults [1.6, 1.7]. For example, in Fig. 1.2, faults 3 and 4 create stuck-on transistors faults. As a further example we consider Fig. 1.3, which represents CMOS implementation of the Boolean function

$$Z = \overline{(A + B)(C + D) + EF}.$$

Two possible shorts numbered 1 and 2 and two possible opens numbered 3 and 4 are indicated in the diagram. Short number 1 can be modeled by s-a-1 of input E; open number 3 can be modeled by s-a-0 of input E or input F or both. On the other hand, short number 2 and open number 4 cannot be modeled by any stuck-at fault because they involve a modification of the network function. For example, in the presence of short number 2 the network function will change to

$$Z = \overline{(A + C)(B + D)(E + F)},$$

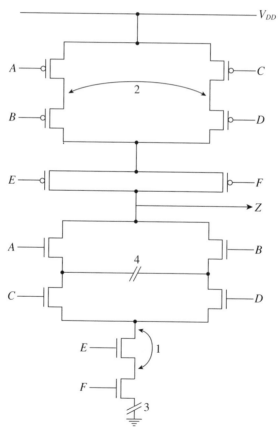

Figure 1.3   CMOS implementation of $Z = \overline{(A + B)(C + D) + EF}$

and open number 4 will change the function

$$Z = \overline{(AC + BD) \cdot EF}.$$

For this reason, a perfect short between the output of the two gates (Fig. 1.4) cannot be modeled by any stuck-at fault. Without a short, the outputs of the gates Z_1 and Z_2 are

$$Z_1 = \overline{AB} \quad \text{and} \quad Z_2 = \overline{CD},$$

whereas with the short,

$$Z_1 = Z_2 = \overline{AB + CD}.$$

1.2.2 BRIDGING FAULTS

With the increase in the number or devices on the VLSI chips, the probability of shorts between two or more signal lines has been significantly increased. Unintended shorts between the lines form a class of permanent faults, known as *bridging faults*, which cannot be modeled as stuck-at faults. It has been observed that physical defects in MOS (Metal Oxide Semiconductor) circuits are manifested as bridging faults more than as any other type of fault [1.8].

Bridging faults can be categorized into three groups:

Input bridging
Feedback bridging
Nonfeedback bridging

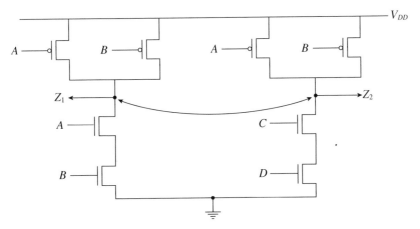

Figure 1.4 Short between outputs Z_1 and Z_2

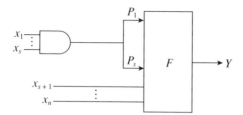

Figure 1.5   Logical model of input bridging $(x_1, \ldots, x_s)$ (from M. Karpovsky and S.Y.H. Su, "Detection and location of input and feedback bridging faults among input and output lines," IEEE Trans. Comput., June 1980. Copyright © 1980 IEEE. Reprinted with permission).

An input bridging fault corresponds to the shorting of a certain number of primary input lines (Fig. 1.5), whereas a feedback bridging fault occurs if there is a short between an output line to an input line (Fig. 1.6). A nonfeedback bridging fault identifies a bridging fault that does not belong to either of the two previous categories. From these definitions, it will be clear that the probability of two lines getting bridged is higher if they are physically close to each other. Most of the existing literature on the bridging faults assumes that the probability of more than two lines shorting together is very low, and *wired logic* is performed at the connections. In general, a bridging fault in positive logic is assumed to behave as a wired-AND (where 0 is the dominant logic value), and a bridging fault in negative logic behaves as a wired-OR (where 1 is the dominant logic value). If bridging between any s lines in a circuit are considered, the number of single bridging faults alone will be $(n/s)!$ and the number of multiple bridging faults will be very much larger.

The presence of a feedback bridging fault can cause a circuit to oscillate or convert it into a sequential circuit. For example, a circuit implementing the func-

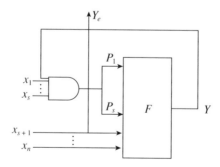

Figure 1.6   Logical model of feedback bridging $(Yx_1, \ldots, x_s)$ (*courtesy of IEEE,* © *1980*)

tion $F(x_1, \ldots, x_i, \ldots, x_n)$ oscillates under feedback bridging $(Yx_1, x_2, \ldots, x_s)$ shown in Fig. 1.6, if the input combination $(x_1, \ldots, x_n)$ satisfies the following condition [1.9]:

$$x_1x_2 \cdots x_sF(0, 0, \ldots, 0, x_{s-1}, \ldots, x_n)\overline{F}(1, 1, \ldots,1, x_{s-1}, \ldots, x_n) = 1. \tag{1.1}$$

For example, if the network of Fig. 1.7(a) has the feedback bridging fault Yx_1x_2 (Fig. 1.7(b)), it will oscillate for the input combination

$$(x_1, x_2, x_3, x_4, x_5, x_6) = (1, 1, 1, 1, 0, 0),$$

because condition (1.1) is satisfied for this input pattern.

As mentioned previously, early work on bridging faults assumed a wired-AND or wired-OR at the connection of the shorted signal lines. Although this assumption is correct for TTL (Transistor Transistor Logic) circuits, it has been realized in recent years that bridging faults in CMOS circuits show characteristics that cannot be represented by the wired logic concept. The effect of such faults has to be analyzed at the transistor level rather than at the gate level. However, the number of possible bridging faults in a circuit at the transistor level can be enormous. Therefore, for practical reasons, only the effect of single bridging fault in a circuit can be considered. It is important to realize that bridging faults are layout dependent, that is, shorts may only occur between signal lines that are adjacent or overlapping in the layout. A short in the transistor-level diagram of the circuit under test may not correspond to an actual bridging fault in the layout of the circuit.

Let us consider a two-input CMOS NOR gate circuit to analyze the effect of

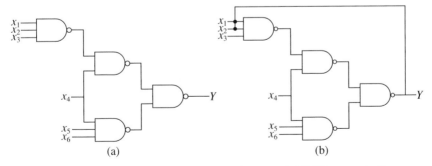

(a)   (b)

Figure 1.7 A network with feedback bridging faults (a) A NAND network, $Y = \overline{x_1x_2x_3}x_4 + x_4x_5x_6$; (b) feedback bridging Yx_1x_2 (from M. Karpovsky and S.Y.H. Su, ''Detection and location of input and feedback bridging faults among input and output lines,'' IEEE Trans. Comput., June 1980. Copyright © 1980 IEEE. Reprinted with permission).

single bridging faults; the circuit and its layout are shown in Figs. 1.8(a) and (b), respectively. The probable bridging fault in this circuit (or in any other CMOS circuit) can be grouped into four categories [1.10]:

1. Metal polysilicon short (*a* in Figs. 1.8(a) and (b))
2. Polysilicon *n*-diffusion short (*c,d* in Figs. 1.8(a) and (b))
3. Polysilicon *p*-diffusion short (*e,f* in Figs. 1.8(a) and (b))
4. Metal polysilicon short (*g* in Figs. 1.8(a) and (b))

Table 1.1 shows the truth table of the NOR gate in the presence of a bridging fault from each of these four categories. It is often assumed that a short creating

Table 1.1   **Expected and Actual NOR Gate Output in Presence of Assumed Bridging Faults**

Input	Output				
A *B*	*Expected*	*Short a*	*Short g*	*Short c*	*Short e*
0 0	1	1	0	0	0
0 1	0	0	0	1	0
1 0	0	1	1	0	0
1 1	0	0	1	1	0

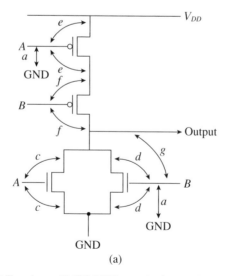

(a)

Figure 1.8   (a) Two-input CMOS NOR gate (points on layout: *a, c, d, e, f, g*);

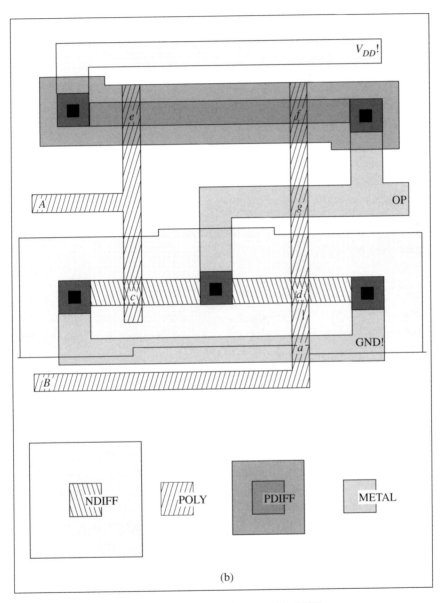

Figure 1.8 continued (b) layout of the NOR gate

a bridging fault is *perfect*, that is, the resistance of the short is close to zero. However, it has been observed that the short resistance can vary from a few ohms to about 5 Kohms [1.11].

1.2.3 BREAKS AND TRANSISTOR STUCK-ON/-OPEN FAULTS IN CMOS

As discussed previously, not all defects in CMOS VLSI can be represented by using the stuck-at fault model. Recent research indicates that breaks and transistor stuck-ons are two other types of defects that, like bridging, may remain undetected if testing is performed based on the stuck-at fault assumption. These defects have been found to constitute a significant percentage of defects occurring in CMOS circuits [1.12]. In the following two subsections, we discuss the effects of these defects on CMOS circuits.

Breaks

Breaks or opens in CMOS circuits are caused either by missing conducting material or extra insulating material. Breaks can be either of the following two types [1.13]:

1. Intragate breaks
2. Signal line breaks

An intragate break occurs internal to a gate. Such a break can disconnect the source, the drain, or the gate from a transistor, identified by b_1, b_2, and b_3, respectively, in Fig. 1.9. The presence of b_3 will have no logical effect on the operation of a circuit, but it will increase the propagation delay; that is, the break will result in a delay fault. Similarly, the break at b_1 will also produce a delay fault without changing the function of the circuit. However, the break at b_2 will make the p-transistor nonconducting; that is, the transistor can be assumed to be *stuck open*.

An intragate break can also disconnect the p-network or the n-network or both networks (b_4, b_5, b_6 in Fig. 1.9) from the circuit. The presence of b_4 or b_5 will have the same effect as the output node getting stuck-at-0 or stuck-at-1, respectively. In the presence of b_6, the output voltage may have an intermittent stuck-at-1 or stuck-at-0 value; thus, if the output node simultaneously drives a p-transistor and an n-transistor, then one of the transistors will be ON for some unpredictable period of time. Signal line breaks can force the gates of transistors in static CMOS circuits to float. As shown in Fig. 1.9, such a break can make the gate of only a p-transistor or an n-transistor to float. It is also possible, depending

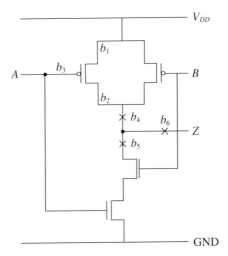

Figure 1.9 Two-input CMOS NAND gate showing occurrence of breaks

on the position of a break, that the gates of both transistors may float, in which case one transistor may conduct and the other remain in a nonconducting state [1.5]. In general, this type of break can be modeled as a stuck-at fault. On the other hand, if two transistors with floating gates are permanently conducting, one of them can be considered as stuck on. If a transistor with a floating gate remains in a nonconducting state due to a signal line break, the circuit will behave in a similar fashion as it does in the presence of the intragate break b_2.

Stuck-On and Stuck-Open Faults

To ensure realistic modeling, faults should be considered at the transistor model, because only at this level is the complete circuit structure known. In other words, circuits should be tested for *shorts* and *opens* at the transistor level. A short corresponds to a stuck-on transistor, whereas an open corresponds to a stuck-open transistor.

A stuck-on transistor fault implies the permanent closing of the path between the source and the drain of the transistor. Although the stuck-on transistor in practice behaves in a similar way as a stuck-closed transistor, there is a subtle difference. A stuck-on transistor has the same drain–source resistance as the on-resistance of a fault-free transistor, whereas a stuck-closed transistor exhibits a drain–source resistance that is significantly lower than the normal on-resistance. In other words: In the case of stuck-closed transistor, the short between the drain and the source is almost perfect, and this is not true for a stuck-on transistor. A transistor stuck-on (stuck-closed) fault may be modeled as a bridging fault from the source to the drain of a transistor. It has been estimated that 10–13% of all faults occurring in CMOS circuits are stuck-on transistor faults [1.12].

A stuck-open transistor implies the permanent opening of the connection between the source and the drain of a transistor. The drain–source resistance of a stuck-open transistor is significantly higher than the off-resistance of a nonfaulty transistor. If the drain–source resistance of a faulty transistor is approximately equal to that of a fault-free transistor, then the transistor is considered to be *stuck off*. For all practical purposes, transistor stuck-off and stuck-open faults are functionally equivalent, and will have the same effect on a CMOS circuit as that of break fault *b2* in Fig. 1.9. Although only about 1% of the CMOS faults are due to stuck-off/stuck-open transistors [1.12], considerable research has been directed at detecting these faults [1.14–1.17]. This is because apart from bridging faults, these are the only faults that can turn a combinational circuit into a sequential circuit [1.18].

Figure 1.10 shows a two-input CMOS NOR gate. A stuck-open fault causes the output to be connected neither to GND nor to V_{DD}. If, for example, transistor *T2* is open-circuited, then for input $AB = 00$, the pull-up circuit will not be active and there will be no change in the output voltage. In fact, the output retains its previous logic state; however, the length of time the state is retained is determined by the leakage current at the output node. Table 1.2 shows the truth table for the two-input CMOS NOR gate. The fault-free output is shown in column Z; the three columns to the right represent the outputs in presence of the three stuck-open (s-op) faults. The first, *As-op*, is caused by any input, drain, or source missing connection to the pull-down FET *T3*. The second, *Bs-op*, is caused by any

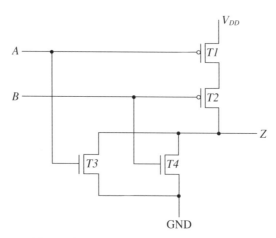

Figure 1.10   A two-input CMOS NOR gate

Table 1.2   **Truth Table for Two-Input CMOS NOR Gate**

A	B	Z	Z (As-op)	Z (Bs-op)	Z (V_{DD}s-op)
0	0	1	1	1	Z_t
0	1	0	0	Z_t	0
1	0	0	Z_t	0	0
1	1	0	0	0	0

input, drain or source missing connection to the pull-down FET *T4*. The third, V_{DD}s-op, is caused by an open anywhere in the series, *p*-channel pull-up connection to V_{DD}. The symbol Z_t is used to indicate that the output state retains the previous logic value. The modeling of the stuck-open faults has been proposed by Wadsack [1.18].

1.2.4 DELAY FAULTS

As mentioned previously, not all manufacturing defects in VLSI circuits can be represented by the stuck-at fault model. The size of a defect determines whether the defect will affect the logic function of a circuit. Smaller defects, which are likely to cause partial open or short in a circuit, have a higher probability of occurrence due to the statistical variations in the manufacturing process [1.19]. These defects result in the failure of a circuit to meet its timing specifications without any alteration of the logic function of the circuit. A small defect may delay the transition of a signal on a line either from 0 to 1, or vice versa. This type of malfunction is modeled by a delay fault.

Two types of delay faults have been proposed in literature:

Gate delay fault [1.20–1.22]
Path delay fault [1.23–1.25]

Gate delay faults have been used to model defects that cause the actual propagation delay of a faulty gate to exceed its specified worst case value. For example, if the specified worst case propagation delay of a gate is *x* units, and the actual delay is $x + \Delta x$ units, then the gate is said to have a delay fault of size Δx. The main deficiency of the gate delay fault model is that it can only be used to model isolated defects, not distributed defects, for example, several small delay defects. The path delay fault model can be used to model isolated as well as distributed defects. In this model, a fault is assumed to have occurred if the propagation delay along a path in the circuit under test exceeds the specified limit.

1.3   Temporary Faults

A major portion of digital system malfunctions are caused by temporary faults [1.26, 1.27]. These faults have also been found to account for more than 90% of the total maintenance expense, because they are difficult to detect and isolate [1.28, 1.29].

In the literature, temporary faults have often been referred to as *intermittent* or *transient* faults with the same meaning. It is only recently that a distinction between the two types of faults has been made [1.30].

Transient faults are nonrecurring temporary faults. They are usually caused by α-particle radiation or power supply fluctuation, and they are not repairable because there is no physical damage to the hardware. They are the major source of failures in semiconductor memory chips.

Intermittent faults are recurring faults that reappear on a regular basis. Such faults can occur due to loose connections, partially defective components, or poor designs. Intermittent faults occurring due to deteriorating or aging components may eventually become permanent. Some intermittent faults also occur due to environmental conditions such as temperature, humidity, vibration, and so forth. The likelihood of such intermittents depends on how well the system is protected from its physical environment through shielding, filtering, cooling, and so on. An intermittent fault in a circuit causes a malfunction of the circuit only if it is active; if it is inactive, the circuit operates correctly. A circuit is said to be in a *fault-active state* if a fault present in the circuit is active, and it is said to be in the *fault-not-active state* if a fault is present but inactive [1.31].

Because intermittent faults are random, they can be modeled only by using probabilistic methods. Several probabilistic models for representing the behavior of intermittent faults have been presented in literature. The first model is a two-state first-order Markov model (see Appendix) presented by Breuer for a specific class of intermittent faults, which are *well behaved* and *signal independent* [1.32]. An intermittent fault is *well behaved* if, during an application of a test pattern, the circuit under test behaves as if either it is fault-free or a permanent fault exists. An intermittent fault is *signal independent* if its being active does not depend on the inputs or the present state of a circuit. Figure 1.11 shows the fault model proposed by Breuer. It assumes that the fault oscillates between the fault-active state (FA) and the fault-not-active state (FN). The transition probabilities indicated in Fig. 1.11 depend on a selected time-step; they have to be changed if this time-step is changed. Lala and Hopkins [1.33] used an adaptation of the Breuer's model that characterizes the transition between the fault-active state and the fault-not-active state by two parameters α and δ, referred to as

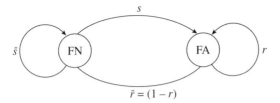

$$\bar{r} = (1 - r)$$

Figure 1.11 Two-state Markov model

frequencies of transition. The ratio α/δ is called the *latency factor*; the higher the latency factor, the lower is the probability of the fault being active.

Kamal and Page [1.34] introduced a zero-order Markov model for intermittent faults, and they suggested a procedure for the detection of a single, well-behaved, signal-independent, intermittent fault in nonredundant combinational circuits. The model assumes prior estimation of the probability that a circuit possesses an intermittent fault and the conditional probability of a fault being active, given that it is present. The fault detection procedure employs the repeated application of tests that are generated to detect permanent faults. After applying a test, the probability of detection of a given intermittent fault is calculated using Bayes' rule [1.35]. This probability approaches 1 if the test is repeated an infinite number of times. However, a finite number of repetitions can be found by using one of the two decision rules. One decision rule is to terminate repetition when the posterior probability (i.e., the probability of a given intermittent fault being present in a circuit after the application of the test) goes below a certain value. The other decision rule is to stop the application of the test when the ''likelihood ratio'' (which is a function of the posterior probabilities) becomes less than a threshold number. Usually the number of repetitions required is still very large. The zero-order Markov model has also been used by Savir [1.36], and by Koren and Kohavi [1.37] for describing the behavior of intermittent faults.

Su *et al.* [1.38] have presented a continuous-parameter Markov model for intermittent faults; this is a generalization of the discrete-parameter model proposed by Breuer. In this model, shown in Fig. 1.12, the transition probabilities depend linearly on the time-step Δt. For example, if a circuit is in the fault-not-active (FN) state at time t, the probability that it will go to the fault-active (FA) state at time $t + \Delta t$ is proportional to Δt. If a constant of proportionality λ is assumed, then this probability is given by $\lambda \Delta t$. Similarly the probability for going from FA state at time t to FN state at time $t + \Delta t$ is $\mu \Delta t$. The time-period during which a circuit stays in state FA (FN) is exponentially distributed with mean $1/\mu$ $(1/\lambda)$. When the time-step Δt is very large, the continuous Markov model reduces

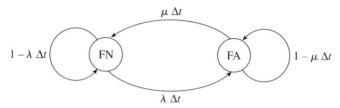

Figure 1.12   Continuous-parameter Markov model (from Su *et al.*, "A continuous parameter Markov model and detection procedures for intermittent faults,"
IEEE Trans. Comput., June 1978. Copyright © 1978 IEEE. Reprinted
with permission).

to the discrete zero-order Markov model, in which case the probability of the
fault being active is $\lambda/(\lambda + \mu)$.

The major problem with the intermittent fault models discussed so far is that
it is very hard to obtain the statistical data needed to verify their validity. The
model proposed by Stifler [1.39] goes a long way to alleviate this problem. It
consists of five states (Fig. 1.13). States *A* and *B* are the fault-active and the
benign state, respectively. If a fault occurs, the error state *E* is entered. *D* is a
fault-detected state, and *F* is a failed state resulting from the propagation of an
undetected error. $\alpha(t)$ represents the probability of transition from the fault-active
to the benign state. $\beta(t)$ is the rate of occurrence of the transition from the benign
state to the fault-active state. $\rho(t)$, $\gamma(t$, and $\epsilon(t)$ denote respectively the rates of
occurrence of error generation, fault detection, and error propagation. Each of
these transition functions is a function only of the time *t*, spent in the source state.
The parameter *C* represents the *coverage* probability, which is the probability of
detecting an error before it causes any damage.

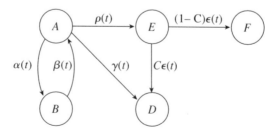

Figure 1.13   Intermittent fault model (from J. I. Stifler, "Robust detection of intermittent faults," Proc. Int. Symp. Fault-tolerant Computing, 1980. Copyright
© 1980 IEEE. Reprinted with permission).

Considerable work has been done in recent years on the *diagnosis* of permanent faults in hardware (see Chap. 2); however, the *diagnosis* of temporary faults remains a major problem. Currently, two types of technique are used to prevent temporary faults from causing system failures: fault masking and concurrent fault detection. The fault-masking techniques tolerate the presence of faults and provide continuous system operation. Concurrent fault detection techniques use *totally self-checking circuits* to signal the presence of faults but not mask them.

References

1.1   Anderson, T., and P. Lee, *Fault-Tolerance: Principles and Practice*, Prentice-Hall International (1981).

1.2   Avizienis, A., "Fault-tolerant systems," *IEEE Trans. Comput.*, 1304–1311 (December 1976).

1.3   Faulkner, T. L., C. W. Bartlett and M. Small, "Hardware design faults: A classification and some measurements," *ICL Technical Jour.*, 218–228 (November 1982).

1.4   Shoji, M., *CMOS Digital Circuit Technology*, Prentice-Hall, (1988).

1.5   Maly, W., P. Nag, and P. Nigh, "Testing oriented analysis of CMOS ICs with opens," *Proc. Intl: Conf. on CAD*, 344–347 (1988).

1.6   Maly, W., "Realistic fault modeling for VLSI testing," *Proc. 24th ACM/ IEEE Design Automation Conf.*, pp. 173–180 (1987).

1.7   Ferguson, J. and J. Shen, "A CMOS fault extractor for inductive fault analysis," *IEEE Trans. on CAD*, pp. 1181–1194 (November 1988).

1.8   Gailay, J., Y. Crouzet, M. Vergniault, "Physical versus logical fault models in MOS LSI circuits: Impact on their testability," *IEEE Trans. Comput.*, 1286–1293 (June 1980).

1.9   Karpovsky, M., and S.Y.H. Su, "Detection and location of input and feedback bridging faults among input and output lines," *IEEE Trans. Comput.*, 523–527 (June 1980).

1.10 Rome Air Development Center, Tech. Report RADC-TR-88-79, "Fault model development for fault tolerant VLSI design." (May 1988).

1.11 Soden, J. M., "Test consideration for gate oxide shorts in CMOS ICs," *IEEE Design and Test*, 56–64 (August 1986).

1.12 Ferguson, J., and J. Shen, "Extraction and simulation of realistic CMOS faults using inductive fault analysis," *Proc. Intl. Test Conf.*, 475–484 (1988).

1.13 Ferguson, J., M. Taylor, and T. Larrabee, "Testing for parametric faults in static CMOS circuits," *Proc. Intl. Test Conf.*, 436–442 (1990).

1.14 El-ziq, Y. M., and R. Cloutier, "Functional-level test generation for stuck-open in CMOS VLSI," *Proc. Intl. Test Conf.*, pp. 536–546 (1981).

1.15 Jha, N. K., "Multiple stuck-open fault detection in CMOS logic circuits," *IEEE Trans. Comput.*, pp. 426–432 (April 1988).

1.16 Koeppe, S., "Optimal layout to avoid CMOS stuck open faults," *Proc. IEEE Design Automation Conf.*, pp. 829–835 (1987).

1.17 Liu, D. L., and E. J. McCluskey, "Designing CMOS circuits for switch-level testability," *IEEE Design and Test of Computers*, pp. 42–49 (August 1987).

1.18 Wadsack, R. L., "Fault modelling and logic simulation of CMOS and MOS integrated circuits," *Bell Syst. Tech. Jour.*, pp. 1149–1475 (May–June 1978).

1.19 David, M. W., "An optimized delay testing technique for LSSD-based VLSI logic circuits," *IEEE VLSI Test Symp.*, pp. 239–246 (1991).

1.20 Hsieh, E. P., R. A. Rasmussen, L. J. Vidunas, and W. T. Davis, "Delay test generation," *Proc. 14th Design Automation Conf.*, pp. 486–491 (June 1977).

1.21 Liaw, C. C., Y. H. Su, and Y. K. Malaiya, "Test generation for delay faults using stuck-at-fault test set," *Proc. Intl. Test Conf.*, pp. 167–175 (1980).

1.22 Kishida, K., F. Shirotori, Y. Ikemoto, S. Ishiyama, and T. Hayashi, "A delay test system for high speed logic LSI's," *Proc. 23rd Design Automation Conf.*, pp. 786–790 (July 1986).

1.23 Smith, G. L., "Model for delay faults based upon path," *Proc. Intl. Test Conf.*, pp. 342–349 (1985).

1.24 Lin, C. J., and S. M. Reddy, "On delay fault testing in logic circuits," *Proc. Intl. Conf. on CAD*, pp. 148–151 (1986).

1.25 Savir, J., and W. H. McAnney, "Random pattern testability of delay fault," *Proc. Intl. Test Conf.*, pp. 163–173 (1986).

1.26 Ball, M., and F. Hardie, "Effects and detection of intermittent failures in digital systems," *Proc. Fall Joint Comput. Conf.*, pp. 329–335 (1969).

1.27 Tasar, O., and V. Tasar, "A study of intermittent faults in digital computers," *AFIPS Conf. Proc.*, pp. 807–811 (1977).

1.28 Clary, J. B., and R. A. Sacane, "Self-testing computers," *IEEE Computer*, 49–59 (October 1979).

1.29 Lala, P. K., and J. I. Missen, "Method for the diagnosis of a single intermittent fault in combinatorial logic circuits," *IEE Jour. Computers and Digital Techniques*, 187–190 (October 1979).

1.30 McCluskey, E. J., and J. F. Wakerly, "A circuit for detecting and analyzing temporary failures," *Proc. COMPCON*, 317–321 (1981).

1.31 Malaiya, Y. K., and S.Y.H. Su, "A survey of methods for intermittent fault analysis," *Proc. Nat. Comput. Conf.*, 577–584 (1979).

1.32 Breuer, M. A., "Testing for intermittent faults in digital circuits," *IEEE Trans. Comput.*, 241–245 (March 1973).

1.33 Lala, J. H., and A. L. Hopkins, "Survival and dispatch probability models for the FTMP computer," *Proc. Intl. Symp. Fault-Tolerant Computing*, 37–43 (1978).

1.34 Kamal, S., and C. V. Page, "Intermittent faults: A model and a detection procedure," *IEEE Trans. Comput.*, 241–245 (July 1974).

1.35 Smith, D. J., *Reliability Engineering*, Pitman (1972).

1.36 Savir, J., "Optimal random testing of single intermittent failures in combinational circuits," *Proc. Intl. Symp. Fault-Tolerant Computing*, 180–185 (1977).

1.37 Koren, I., and Z. Kohavi, "Diagnosis of intermittent faults in combinational networks," *IEEE Trans. Comput.*, 1154–1158 (November 1977).

1.38 Su, S.Y.H., I. Koren, and Y. K. Malaiya, "A continuous parameter Markov model and detection procedures for intermittent faults," *IEEE Trans. Comput.*, 567–569 (June 1978).

1.39 Stifler, J. I., "Robust detection of intermittent faults," *Proc. Intl. Symp. Fault-Tolerant Computing*, 216–218 (1980).

Chapter 2 | Test Generation for Combinational Logic Circuits

2.1 Fault Diagnosis of Digital Circuits

Digital circuits, even when designed with highly reliable components, do not operate forever without developing some faults. When a circuit ultimately does develop a fault, it has to be detected and located so that its effect can be removed. *Fault detection* means the discovery of something wrong in a digital system or circuit. *Fault location* means the identification of the faults with components, functional modules, or subsystems, depending on the requirements. *Fault diagnosis* includes both fault detection and fault location.

Fault detection in a logic circuit is carried out by applying a sequence of test inputs and observing the resulting outputs. Therefore, the cost of testing includes the generation of test sequences and their application. One of the main objectives in testing is to minimize the length of the test sequence. Any fault in a nonredundent* n-input combinational circuit can be completely tested by applying all 2^n input combinations to it; however, 2^n increases very rapidly as n increases. For a sequential circuit with n inputs and m flip-flops, the total number of input combinations necessary to test the circuit exhaustively is $2^n \times 2^m = 2^{m+n}$. If, for example, $n = 20$, $m = 40$, there would be 2^{60} tests. At a rate of 10,000 tests per second, the total test time for the circuit would be about $3\frac{1}{2}$ million years! Fortunately, a complete truth-table exercise of a logic circuit is not necessary—only the number of input combinations that detect most of the faults in the circuit is required. To determine which faults have been detected by a set of test patterns,

---

*A circuit is said to be nonredundant if the function realized by the circuit is not the same as the function realized by the circuit in the presence of a fault.

a process known as *fault-simulation* is performed [2.1]. For example, if a circuit has x stuck-at faults, then fault simulation is the process of applying every test pattern to the fault-free circuit, and to each of the x copies of the circuit containing exactly one stuck-at fault. When all test patterns have been simulated against all the faults x, the detected faults f are used to compute the *fault coverage* (f_c) which is defined as

$$f_c = \frac{f}{x}.$$

Instead of simulating one fault at a time, the method of *parallel simulation,* which uses the word size N of a host computer to process N faults at a time, can be employed. Another approach is to utilize the *deductive simulation* method, which allows simulation of all faults simultaneously but is much harder to implement and requires enormous memory capacity for larger circuits.

It has been shown that the time required to compute test patterns for a combinational circuit grows in proportion to the square of the number of gates in the circuit [2.2]. Hence, for circuits of VLSI complexity the computation time required for test generation is often unacceptably high.

2.2   Test Generation Techniques for Combinational Circuits

There are several methods available for deriving tests for combinational circuits. All these methods are based on the assumption that the circuit under test is non-redundant and only a single stuck-at fault is present at any time. In this section, we discuss each method in some detail.

2.2.1   ONE-DIMENSIONAL PATH SENSITIZATION

The basic principle involved in *path sensitizing* is to choose some path from the origin of the failure to the circuit output. The path is said to be ''sensitized'' if the inputs to the gates along the path are assigned values so as to propagate the fault along the chosen path to the output [2.3].

The method can be illustrated with an example. Let us consider the circuit shown in Fig. 2.1 and suppose that the fault is line X_3 s-a-1. To test X_3 for s-a-1, X_5 and G_2 must be set at 1 and X_3 set at 0, so that $G_5 = 1$ if the fault is absent. We now have a choice of propagating the fault from G_5 to the circuit output Z via a path through $G_7 G_9$ or through $G_8 G_9$. To propagate through $G_7 G_9$ requires the output of G_4 and G_8 to be 1. If $G_4 = 1$, the output of G_7 depends on the

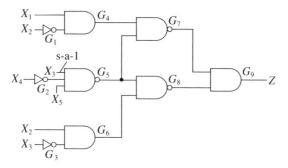

Figure 2.1 Circuit with a s-a-1 fault

output of G_5, and similarly if $G_8 = 1$, the circuit output depends on G_7 only. This process of propagating the effect of the fault from its original location to the circuit output is known as the *forward trace*. The next phase of the method is the *backward trace,* in which the necessary gate conditions to propagate the fault along the sensitized path are established. For example, to set G_4 at 1 both X_1 and G_1 must be set at 1; $G_1 = 1$ implies $X_2 = 0$. For G_2 to be 1, X_4 must be set at 0. for G_8 to be 1, G_6 must be 0, which requires either $X_2 = 0$ or $G_3 = 0$. Since X_2 has already been specified as 0, the output of G_6 will be 0. It is worth noting that G_6 cannot be set at 0 by making $G_3 = 0$, because this would imply $X_3 = 1$, which is inconsistent with the previous assignment of X_3. Therefore, the test $X_1X_2X_3X_4X_5 = 10001$ detects the fault X_3 s-a-1, because the output of Z will be 0 for the fault-free circuit and 1 for the circuit having the fault.

In general, the input combination generated by the path-sensitization procedure for propagating a fault to the output may not be unique. For example, the X_3 s-a-1 fault in the circuit of Fig. 2.1 can also be detected by the test $X_1X_2X_3X_4X_5 = -1001$; this is done by sensitizing the path $G_5G_8G_9$. X_1 has an unspecified value in this test; that is, the test is independent of input X_1.

The flaw in the one-dimensional path sensitization technique is that only one path is sensitized at a time. The following example shows the inadequacy of this procedure [2.4].

Let us try to derive a test for the fault G_2 s-a-0 in Fig. 2.2 by sensitizing the path $G_2G_6G_8$. We set $X_2 = X_3 = 0$ so as to propagate the effect of the fault through gate G_2. Setting $X_4 = 0$ propagates it through gate G_6. To propagate it through gate G_8, we require $G_4 = G_5 = G_7 = 0$. Because X_2 and X_4 have already been set to 0, we have $G_3 = 1$, which makes $G_7 = 0$. To make $G_5 = 0$, X_1 must be set to 1; consequently, $G_1 = 0$, which with $X_2 = 0$ would make $G_4 = 1$. Therefore, we are unable to propagate through G_8. Similarly, we can show that it is impossible to sensitize the single path $G_2G_5G_8$. However, we note that $X_1 =$

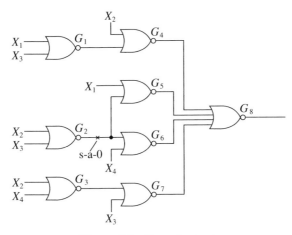

Figure 2.2   Example circuit

$X_4 = 0$ sensitizes the two paths simultaneously and also makes $G_4 = G_7 = 0$. Thus, two inputs to G_8 change from 0 to 1 as a result of the fault G_2 s-a-0, while the remaining two inputs remain fixed at 0. The fault will cause the output of G_8 to change from 1 to 0 and $\overline{X}_1\overline{X}_2\overline{X}_3\overline{X}_4$ is a test for G_8 s-a-0.

The foregoing example shows the necessity of sensitizing more than one path in deriving tests for certain faults. This is the principal idea behind the *D-algorithm* (see Sec. 2.2.3).

2.2.2   BOOLEAN DIFFERENCE

The basic principle of the Boolean difference is to derive two Boolean expressions—one of which represents normal fault-free behavior of the circuit, and the other, the logical behavior under an assumed single s-a-1 or s-a-0 fault condition. These two expressions are then exclusive-ORed; If the result is 1, a fault is indicated [2.5].

Let $F(X) = F(x_1, \ldots, x_n)$ be a logic function of n variables. If one of the inputs to the logic function, for example, input x_i, is faulty, then the output would be $F(x_1, \ldots, \overline{x}_i, \ldots, x_n)$. The Boolean difference of $F(X)$ with respect to x_i is defined as

$$\frac{dF(x_1, \ldots, x_i, \ldots, x_n)}{dx_i} = \frac{dF(X)}{dx_i}$$

$$= F(x_1, \ldots, x_i, \ldots, x_n) \oplus F(x_1, \ldots, \overline{x}_i, \ldots, x_n).$$

The function $dF(X)/dx_i$ is called the Boolean difference of $F(X)$ with respect to x_i.

It is easy to see that when $F(x_1, \ldots, x_i, \ldots, x_n) \neq F(x_1, \ldots, \bar{x}_i, \ldots, x_n)$, $dF(X)/dx_i = 1$, and that when $F(x_1, \ldots, x_i, \ldots, x_n) = F(x_1, \ldots, x_i, \ldots, x_n)$, $dF(X)/dx_i = 0$. To detect a fault on x_i, it is necessary to find input combinations (tests) so that whenever x_i changes to $\bar{x}_i$ (due to a fault), $F(x_1, \ldots, x_i, \ldots, x_n)$ will be different from $F(x_1, \ldots, \bar{x}_i, \ldots, x_n)$. In other words, the aim is to find input combinations for each fault occurring on x_i such that $dF(X)/dx_i = 1$.

Some useful properties of the Boolean difference are

1. $\dfrac{d\overline{F(X)}}{dx_i} = \dfrac{dF(X)}{dx_i}$; $\overline{F(X)}$ denotes the complement of $F(X)$.

2. $\dfrac{dF(X)}{dx_i} = \dfrac{dF(X)}{d\bar{x}_i}$.

3. $\dfrac{d}{dx} \cdot \dfrac{dF(X)}{dx_i} = \dfrac{d}{dx} \dfrac{dF(X)}{dx_i}$.

4. $\dfrac{d[F(X)G(X)]}{dx_i} = F(X) \dfrac{dG(X)}{dx_i} \oplus G(X) \dfrac{dF(X)}{dx_i} \oplus \dfrac{dF(X)}{dx_i} \cdot \dfrac{dG(X)}{dx_i}$.

5. $\dfrac{d[F(X) + G(X)]}{dx_i} = \overline{F(X)} \dfrac{dG(X)}{dx_i} \oplus \overline{G(X)} \dfrac{dF(X)}{dx_i} \oplus \dfrac{dF(X)}{dx_i} \cdot \dfrac{dG(X)}{dx_i}$.

A Boolean function $F(X)$ is said to be *independent of* x_i if and only if $F(X)$ is logically invariant under complementation of x_i, that is, if

$$F(x_1, \ldots, x_i, \ldots, x_n) = F(x_1, \ldots, \bar{x}_i, \ldots, x_n).$$

This implies that a fault in x_i will not affect the final output $F(X)$ and $dF(x)/dx_i = 0$. Some additional properties can now be added to the original set (1–5):

6. $\dfrac{dF(X)}{dx_i} = 0$ if $F(X)$ is independent of x_i.

7. $\dfrac{dF(X)}{dx_i} = 1$ if $F(X)$ depends only on x_i.

8. $\dfrac{d[F(X)G(X)]}{dx_i} = F(X) \dfrac{dG(X)}{dx_i}$ if $F(X)$ is independent of x_i.

9. $\dfrac{d[F(X) + G(X)]}{dx_i} = \overline{F(X)} \dfrac{dG(X)}{dx_i}$ if $F(X)$ is independent of x_i.

To illustrate how the Boolean difference is used, we look at two examples.

Example 1   Consider the logic circuit shown in Fig. 2.3(a). Find the Boolean difference with respect to x_3. We have

$$\frac{dF(X)}{dx_3} = \frac{d(x_1x_2 + x_3x_4)}{dx_3}$$

$$= \overline{x_1x_2}\frac{d(x_3x_4)}{dx_3} \oplus x_3x_4\frac{d(x_1x_2)}{dx_3} \oplus \frac{d(x_1x_2)}{dx_3}\cdot\frac{d(x_3x_4)}{dx_3} \qquad \text{(By property 5)}$$

$$= \overline{x_1x_2}\frac{d(x_3x_4)}{dx_3} \qquad\qquad\qquad\qquad\qquad\qquad \text{(By property 6)}$$

$$= \overline{x_1x_2}\left(x_3\frac{dx_4}{dx_3} \oplus x_4\frac{dx_3}{dx_3} \oplus \frac{dx_3}{dx_3}\cdot\frac{dx_4}{dx_3}\right) \qquad \text{(By property 4)}$$

$$= \overline{x_1x_2}x_4. \qquad\qquad\qquad\qquad\qquad\qquad \text{(By properties 6 and 7)}$$

This means that a fault on x_3 will cause the output to be in error only if $\overline{x_1x_2}x_4 = 1$, that is, if x_1 or x_2 (or both) are equal to 0 and x_4 is equal to 1. This can be verified by inspection of Fig. 2.3(a).

Example 2 Consider the logic circuit shown in Fig. 2.3(b). Find the Boolean difference with respect to x_2. We have

$$\frac{dF(X)}{dx_2} = \frac{d(x_1x_2 + x_1\bar{x}_2)}{dx_2}$$

$$= \overline{x_1x_2}\frac{d(x_1\bar{x}_2)}{dx_2} \oplus x_1\bar{x}_2\frac{d(x_1x_2)}{dx_2} \oplus \frac{d(x_1x_2)}{dx_2}\cdot\frac{d(x_1\bar{x}_2)}{dx_2} \qquad \text{(By property 5)}$$

$$= \overline{x_1x_2}x_1 \oplus \overline{x_1\bar{x}_2}\cdot x_1 \oplus x_1 \qquad\qquad \text{(By properties 1, 7, and 8)}$$

$$= x_1(\overline{x_1x_2} \oplus \overline{x_1\bar{x}_2}) \oplus x_1$$

$$= x_1(\overline{x_1x_2}\cdot x_1\bar{x}_2 + \overline{x_1\bar{x}_2}\cdot x_1x_2) \oplus x_1$$

$$= x_1[(\bar{x}_1 + \bar{x}_2)x_1\bar{x}_2 + (\bar{x}_1 + x_2)x_1x_2] \oplus x_1$$

$$= x_1[x_1\bar{x}_2 + x_1x_2] \oplus x_1$$

$$= x_1 \oplus x_1$$

$$= 0.$$

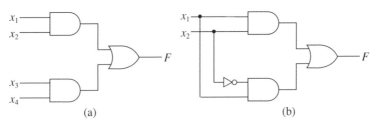

(a)   (b)

Figure 2.3 Circuit examples

This means that a fault in x_2 will not cause the output to be in error, which indicates that the circuit is not really a function of x_2; this can be verified by noting that the original output $F(X) = x_1 \cdot x_2 + x_1 \cdot \overline{x_2} = x_1$.

So far the Boolean difference method has been applied to derive tests for input line faults; it can also be used for faults on lines internal to the circuit.

Let a combinational circuit realize the function $F(X)$, and let h be an internal wire in the circuit. Tests for h can be found by expressing F as a function of h, $F(x_1, x_2, \ldots, x_n, h)$, and h as a function of the inputs $h(x_1, x_2, \ldots, x_n)$.

As an example, consider the circuit of Fig. 2.4 and find tests to detect s-a-0 and s-a-1 faults on h:

$$F = x_1 x_2 + x_3 x_4 + \overline{x_2 x_4}$$
$$= h + \underbrace{(x_3 x_4 + \overline{x_2 x_4})}_{G},$$

$$\frac{dF}{dh} = \overline{G}\,\frac{dh}{dh} \oplus \overline{h}\,\frac{dG}{dh} \oplus \frac{dG}{dh} \cdot \frac{dh}{dh}$$

$$= \overline{(x_3 x_4 + \overline{x_2 x_4})} \cdot \frac{dh}{dh} \oplus \overline{h} \cdot \frac{d(x_3 x_4 + \overline{x_2 x_4})}{dh} \oplus \frac{d(x_3 x_4 + \overline{x_2 x_4})}{dh} \cdot \frac{dh}{dh}$$

$$= \overline{(x_3 x_4 + \overline{x_2 x_4})} \oplus 0 \oplus 0$$

$$= \overline{x_3 x_4} \cdot \overline{\overline{x_2 x_4}} = (\overline{x_3} + \overline{x_4})x_2 x_4 = x_2 \overline{x_3} x_4.$$

Tests for h s-a-0 are given by

$$h \cdot \frac{dF}{dh} = x_1 x_2 \cdot x_2 \overline{x_3} x_4 = x_1 x_2 \overline{x_3} x_4,$$

and h s-a-1 by

$$\overline{h} \cdot \frac{dF}{dh} = \overline{x_1 x_2} \cdot x_2 \overline{x_3} x_4 = (\overline{x_1} + \overline{x_2})x_2 \overline{x_3} x_4 = \overline{x_1} x_2 \overline{x_3} x_4.$$

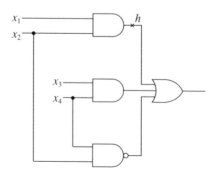

Figure 2.4   Circuit with an internal fault

In general, the conventional Boolean difference is not capable of deriving tests for all the internal nodes of a logic network. This is illustrated by the following example.

Example 3 Given the logic network of Fig. 2.5, determine the tests for checking all single-node faults.

The network function is given by

$$F = \bar{x}_1 x_2 \bar{x}_3 + \bar{x}_4 \bar{x}_2 x_3.$$

The conventional Boolean differences of F with respect to x_1, x_2, and x_3 are

$$\frac{dF}{dx_1} = x_2 \bar{x}_3 + \bar{x}_2 x_3,$$

$$\frac{dF}{dx_2} = \bar{x}_1 \bar{x}_3 + \bar{x}_1 x_3,$$

$$\frac{dF}{dx_3} = \bar{x}_1 x_2 + \bar{x}_1 \bar{x}_2.$$

The test inputs resulting from these Boolean differences will check all input-line faults and are as follows:

x_1 (s-a-0)	1 1 0	or 1 0 1
x_1 (s-a-1)	0 1 0	or 0 0 1
x_2 (s-a-0)	0 1 0	or 0 1 1
x_2 (s-a-1)	0 0 0	or 0 0 1
x_3 (s-a-0)	0 1 1	or 0 0 1
x_3 (s-a-1)	0 1 0	or 0 0 0

One may select the test set (110, 010, 001) as the set of tests capable of detecting the faults on primary input lines. However, this test set may or may not

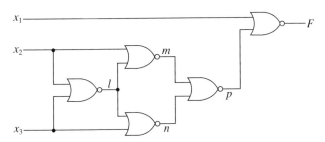

Figure 2.5 A network implementation for $F = \bar{x}_1 x_2 \bar{x}_3 + \bar{x}_1 \bar{x}_2 x_3$

be sufficient to detect all internal-node faults. To develop a complete set of tests that will detect both input line and internal-node faults, it is necessary to find tests that will exercise completely each and every path connecting a primary input to the primary output. This can be carried out by using the *partial* Boolean difference technique [2.6].

In general, a conventional Boolean difference expression may be formed by concatenating individual Boolean differences. For example, if $Z = f(y)$ and $y = f(x)$, then

$$\frac{dZ}{dx} = \frac{dZ}{dy} \cdot \frac{dy}{dx},$$

where $dZ \neq dx$ is termed the partial Boolean difference with respect to x. The partial Boolean difference associated with the path $x_2-l-n-p-F$ in Fig. 2.5 is given by

$$\frac{dF}{dx_2} = \frac{dF}{dp} \cdot \frac{dp}{dn} \cdot \frac{dn}{dl} \cdot \frac{dl}{dx_2}.$$

Because

$$\frac{dF}{dp} = \frac{d}{dp}(\bar{x}_1 \cdot \bar{p}) = \bar{x}_1,$$

$$\frac{dp}{dn} = \frac{d}{dn}(\bar{m} \cdot \bar{n}) = \bar{m} = x_2 + l = x_2 + \bar{x}_2 \cdot \bar{x}_3,$$

$$\frac{dn}{dl} = \frac{d}{dl}(\bar{l}\bar{x}_3) = \bar{x}_3,$$

$$\frac{dl}{dx_2} = \frac{d}{dx_2}(\bar{x}_2 \cdot \bar{x}_3) = \bar{x}_3,$$

it follows that

$$\frac{dF}{dx_2} = \bar{x}_1 \cdot (x_2 + \bar{x}_2\bar{x}_3) \cdot \bar{x}_3 \cdot \bar{x}_3$$

$$= \bar{x}_1 x_2 \cdot \bar{x}_3 + \bar{x}_1\bar{x}_2\bar{x}_3 = \bar{x}_1\bar{x}_3.$$

Therefore, the tests that will exercise the path $x_2-l-n-p-F$ are

$$\bar{x}_1 \cdot x_2 \cdot \bar{x}_3 \quad \text{or} \quad (0 \ 1 \ 0)$$

and

$$\bar{x}_1 \cdot \bar{x}_2 \cdot \bar{x}_3 \quad \text{or} \quad (0 \ 0 \ 0).$$

Proceeding in a similar manner, the partial Boolean difference associated with path $x_3 - n - p - F$ is given by

$$\frac{dF}{dx_3} = \frac{dF}{dp} \cdot \frac{dp}{dn} \cdot \frac{dn}{dx_3}$$

$$= \bar{x}_1 \cdot \bar{m} \cdot \bar{l}$$

$$= \bar{x}_1 \cdot (x_2 + \bar{x}_2 \bar{x}_3) \cdot (x_2 + x_3)$$

$$= \bar{x}_1 x_2,$$

which yields the tests

$$\bar{x}_1 x_2 \bar{x}_3 \qquad (0 \ 1 \ 0)$$

and

$$\bar{x}_1 x_2 x_3 \qquad (0 \ 1 \ 1).$$

The Boolean difference method generates all tests for every fault in a circuit. It is a complete algorithm and does not require any trial and error. However, the method is costly in computation time and memory requirements.

2.2.3 D-ALGORITHM

The D-algorithm is the first algorithmic method for generating tests for nonredundant combinational circuits [2.7]. If a test exists for detecting a fault, the D-algorithm is guaranteed to find this test. Before the D-algorithm can be discussed in detail, certain new terms must be defined.

Singular Cover

The singular cover of a logic gate is basically a compact version of the truth table. Figure 2.6 shows the singular cover for a two-input NOR gate; Xs or blanks are used to denote that the position may be either 0 or 1. Each row in the singular cover is termed a *singular cube*. The singular cover of a network is just the

a	b	c		a	b	c
0	0	1		0	0	1
0	1	0		X	1	0
1	0	0		1	X	0
(a)				(b)		

Figure 2.6 (a) Truth table; (b) singular cover

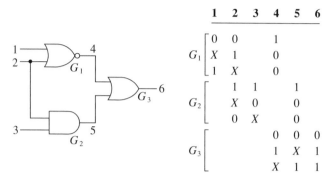

	1	2	3	4	5	6
G_1	0	0		1		
	X	1		0		
	1	X		0		
G_2			1	1		1
			X	0		0
			0	X		0
G_3				0	0	0
				1	X	1
				X	1	1

Figure 2.7 (a) A circuit; (b) singular covers of each gate in the circuit

set of singular covers of each of its gates on separate rows in the table. This is illustrated by the example in Fig. 2.7.

Propagation D-Cubes

D-cubes represent the input–output behavior of the good and the faulty circuit. The symbol D may assume only one value 0 or 1; $\overline{D}$ takes on the value opposite to D; that is, if $D = 1$, $\overline{D} = 0$, and if $D = 0$, $\overline{D} = 1$. The definitions of D and $\overline{D}$ could be interchanged, but they should be consistent throughout the circuit. Thus, all Ds in a circuit imply the same value (0 or 1), and all $\overline{D}$s will have the opposite value.

The *propagation D-cubes* of a gate are those that cause the output of the gate to depend only on one or more of its specified inputs (and hence to propagate a fault on these inputs to the output). The propagation D-cubes for a two-input NOR gate are

a	b	c
0	D	$\overline{D}$
D	0	$\overline{D}$
D	D	$\overline{D}$

The propagation D-cubes $0D\overline{D}$ and $D0\overline{D}$ indicate that if one of the inputs of the NOR gate is 0, the output is the complement of the other input; $DD\overline{D}$ propagates multiple-input changes through the NOR gate.

Propagation D-cubes can be derived from the singular cover, or by inspection. To systematically construct propagation D-cubes, cubes with different output values in a gate's singular cover are intersected, using the following algebraic rules:

$$
\begin{array}{c c c c}
 & a & b & c \\
\hline
C_1 & 1 & 1 & 0 \\
C_2 & X & 0 & 1 \\
C_3 & 0 & X & 1 \\
\end{array}
$$

(a)

$$
\begin{array}{c c c c}
 & a & b & c \\
\hline
C_1 \cap C_2 & 1 & D & \bar{D} \\
C_1 \cap C_3 & D & 1 & \bar{D} \\
\end{array}
$$

(b)

Figure 2.8 (a) Singular covers of the NAND gate; (b) propagation D-cubes of the NAND gate

$$0 \cap 0 = 0 \cap X = X \cap 0 = 0,$$

$$1 \cap 1 = 1 \cap X = X \cap 1 = 1,$$

$$X \cap X = X,$$

$$1 \cap 0 = D,$$

$$0 \cap 1 = \bar{D}.$$

For example, the propagation D-cubes of the two-input NAND gate can be formed from its singular covers, as shown in Fig. 2.8.

Primitive D-Cube of a Fault

The primitive D-cube of a fault is used to specify the existence of a given fault. It consists of an input pattern that brings the influence of a fault to the output of the gate. For example, if the output of the NOR gate shown in Fig. 2.6 is s-a-0, the corresponding primitive D-cube of a fault is

$$
\begin{array}{c c c}
a & b & c \\
\hline
0 & 0 & D \\
\end{array}
$$

Here, the D is interpreted as being a 1 if the circuit if fault-free and a 0 if the fault is present. The primitive D-cubes for the NOR gate output s-a-1 are

$$
\begin{array}{c c c}
a & b & c \\
\hline
1 & X & \bar{D} \\
X & 1 & \bar{D} \\
\end{array}
$$

The primitive D-cube of any fault in a gate can be obtained from the singular covers of the normal and the faulty gates in the following manner:

1. Form the singular covers of the fault-free and the faulty gate. Let α_0 and α_1 be sets of cubes in the singular covers of the fault-free gate the output

coordinates of which are 0 and 1, respectively, and let β_0 and β_1 be the corresponding sets in the singular covers of the faulty gate.

2. Intersect members of α_1 with members of β_0, and members of α_0 with members of β_1. The intersection rules are similar to those used for propagation D-cubes.

The primitive D-cubes of faults obtained from $\alpha_1 \cap \beta_0$ correspond to those inputs that produce a 1 output from the fault-free gate and a 0 output from the faulty gate. The primitive D-cubes of faults obtained from $\alpha_0 \cap \beta_1$ correspond to those inputs that produce a 0 output from the fault-free gate and a 1 output from the faulty gate.

Example 4   Consider a three-input NAND gate with input lines a, b, and c, and output line d. The singular cover for the NAND gate is

	a	b	c	d	
C_{1g}	0	X	X	1	
C_{2g}	X	0	X	1	α_1
C_{3g}	X	X	0	1	
C_{4g}	1	1	1	0	α_0

Assuming the input line b is s-a-1, the singular cover for the faulty NAND gate is

	a	b	c	d	
C_{1f}	0	X	X	1	
C_{2f}	X	X	0	1	β_1
C_{3f}	1	X	1	0	β_0

Therefore,

$$C_{1g} \cap C_{3f} = \overline{D} \quad X \quad 1 \quad \overline{D}, \qquad C_{4g} \cap C_{1f} = D \quad 1 \quad 1 \quad \overline{D},$$

$$C_{2g} \cap C_{3f} = 1 \quad 0 \quad 1 \quad D, \qquad C_{4g} \cap C_{2f} = 1 \quad 1 \quad D \quad \overline{D}.$$

$$C_{3g} \cap C_{3f} = 1 \quad X \quad \overline{D} \quad D,$$

The primitive D-cube of the b s-a-1 fault is $101D$. The primitive D-cubes of all stuck-at faults for the three-input NAND gate are

a	b	c	d	Fault
0	X	X	D	d s-a-0
X	0	X	D	d s-a-0

X	X	0	D	d s-a-0
1	1	1	$\overline{D}$	d s-a-1
0	1	1	D	a s-a-1
1	0	1	D	b s-a-1
1	1	0	D	c s-a-1

D-Intersection

Finally, we need to consider the concept of D-intersection, which provides the tool for building sensitized paths. This is first explained by a simple example. Consider the simple circuit shown in Fig. 2.9. We attempt to generate a test for the 2 s-a-0 fault, described by the D-cube of the fault:

1	2	4
0	1	D

To transmit the $\overline{D}$ on line 4 through G_2, we must try and match, that is, intersect, the $\overline{D}$ specification with one of the propagation D-cubes for G_2. Such a match is possible if we use the propagation D-cube:

3	4	5
0	$\overline{D}$	D

This produces a full circuit D-cube:

1	2	3	4	5
0	1	0	$\overline{D}$	D

Thus, setting $X_1 = 0$, $X_2 = 1$, $X_3 = 0$ will sensitize a path from line 2 through line 4 to line 5, and this will therefore test for inverse polarity faults on these connections.

It is worth noting that intersection of the D-cube of the fault with the other single D-input propagation cube $(\overline{D}0D)$ would not be successful, because in the first cube the status of line 4 is $\overline{D}$, whereas in the second cube it is required to be

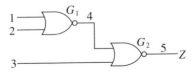

Figure 2.9 Circuit to illustrate D-intersection

set at 0. This is incompatible with the requirement. The full set of rules for the
D-cube intersection is as follows [2.8]:

Let $A = (a_1, a_2, \ldots, a_n)$ and $B = (b_1, b_2, \ldots, b_n)$ be D-cubes where a_i and
b_j equal 0, 1, X, D, or $\overline{D}$ for $i, j = 1, 2, \ldots, n$. The D-intersection, denoted by
$A \cap B$, is given by:

1. $X \cap a_i = a_i$.

2. If $a_i \neq X$ and $b_i \neq X$, then

$$a_i \cap b_i = \begin{cases} a_i & \text{if } b_i = a_i, \\ \varnothing & \text{otherwise.} \end{cases}$$

Finally, $A \cap B = \varnothing$, the empty cube, if for any i, $a_i \cap b_i = \varnothing$; otherwise,

$$A \cap B = a_i \cap b_i, \ldots, a_n \cap b_n.$$

For example,

$$(1X1D0) \cap (X\overline{D}1D0) = 1\overline{D}1D0,$$
$$(01\overline{D}X1) \cap (00XD1) = 0\varnothing\overline{D}D1 = \varnothing.$$

Now that singular cubes, primitive D-cubes of a fault, propagation D-cubes, and
D-intersections have been defined, we shall discuss the D-algorithm in detail.

The first stage of the D-algorithm consists of choosing a primitive D-cube of
the fault under consideration. The next step is to sensitize all possible paths from
the faulty gate to a primary output of the circuit; this is done by successive
intersection of the primitive D-cube of the fault with the propagation D-cubes of
successor gates. The procedure is called the D-*drive*. The D-drive is continued
until a primary output has a D or $\overline{D}$. The final step is the consistency operation,
which is performed to develop a consistent set of primary input values that will
account for all lines set to 0 or 1 during the D-drive.

The application of the D-algorithm is demonstrated in Table 2.1 by deriving
a test for detecting the fault 6 s-a-0 in the circuit of Fig. 2.10. The test for line 6
s-a-0 is 1010.

Fortunately, no inconsistencies were encountered in the foregoing example.
When they are, one must seek a different path for propagating the fault to the
output. This is illustrated by deriving the test for 5 s-a-1 in the circuit diagram of
Fig. 2.11.

The singular covers and the propagation D-cube of the circuit are shown in
Tables 2.2 and 2.3, respectively. The blanks in the tables are treated as Xs while
performing intersections. The D-drive along lines 5–8–9 and the consistency

Table 2.1 **Application of the *D*-Algorithm**

G₁			G₂			G₃			G₄			G₅		
1	2	5	3	4	6	3	5	7	2	6	8	7	8	9
X	1	0	1	0	D	0	X	1	0	D	$\overline{D}$	1	D	$\overline{D}$
1	X	0	0	1	D	X	0	1	D	0	$\overline{D}$	D	1	$\overline{D}$
0	0	1	1	1	$\overline{D}$	1	1	0	D	D	$\overline{D}$	D	D	$\overline{D}$
Singular cover			Primitive *D*-cube of fault			Singular cover			Propagation *D*-cube			Propagation *D*-cube		

	1	2	3	4	5	6	7	8	9
D-drive operation									
1. Select *pdcf* for 6 s-a-0			1	0		D			
2. Intersect with Gate 4 propagation *D*-cube		0	1	0		D		$\overline{D}$	
3. Intersect with Gate 5 propagation *D*-cube		0	1	0		D	1	$\overline{D}$	D
(N.B. G₅ *D*-cube: polarity inverted)									
End of *D*-drive									
Consistency operation									
1. Check line 7 is at 1 from G₃ singular cover		0	1	0	0	D	1	$\overline{D}$	D
Set line 5 at 0									
2. Check line 5 is at 0 from G₁ singular cover;	1	0	1	0	0	D	1	$\overline{D}$	D
set primary input 1 at 1									
End of Consistency									

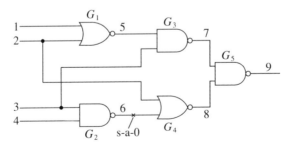

Figure 2.10   Circuit under test

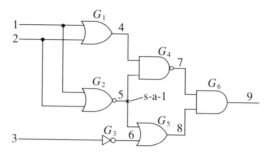

Figure 2.11   Circuit under test

operations are shown in Table 2.4. Any D-cube that represents a partially formed test during the D-drive is called a "test cube" and is represented by tc and a superscript denoting the step at which it is obtained. The primitive D-cube of the fault is chosen as the initial test cube tc^0.

As can be seen from the table the consistency operation terminates unsuccessfully due to the null intersection tc^5. If j is chosen instead of i, tc^3 becomes undefined. We now attempt to D-drive along lines 5–7–9; this is shown in Table

Table 2.2   **Singular Cover**

		1	2	3	4	5	6	7	8	9
Gate 1	a	X	1		1					
	b	1	X		1					
	c	0	0		0					
Gate 2	d	X	1			0				
	e	1	X			0				
	f	0	0			1				
Gate 3	g			0			1			
	h			1			0			
Gate 4	i				0	X		1		
	j				X	0		1		
	k				1	1		0		
Gate 5	l					X	1		1	
	m					1	X		1	
	n					0	0		0	
Gate 6	o							X	0	0
	p							0	X	0
	q							1	1	1

Table 2.3   Propagation D-Cubes

		1	2	3	4	5	6	7	8	9
Gate 1	a_d	0	D		D					
	b_d	D	0		D					
Gate 2	c_d	0	D			$\overline{D}$				
	d_d	D	0			$\overline{D}$				
Gate 3	$\{e_d$			D			$\overline{D}$			
Gate 4	f_d				1	D		$\overline{D}$		
	g_d				D	1		$\overline{D}$		
Gate 5	h_d					0	D		D	
	i_d					D	0		D	
Gate 6	j_d							D	1	D
	k_d							1	D	D

Table 2.4   D-Drive Along Lines 5–8–9 and Consistency Operations

	1	2	3	4	5	6	7	8	9
D-drive									
tc^0	1	0			$\overline{D}$				
$tc^1 = tc^0 \cap i_d$	1	0			$\overline{D}$	0		$\overline{D}$	
$tc^2 = tc^1 \cap k_d$	1	0			$\overline{D}$	0	1	$\overline{D}$	$\overline{D}$
Consistency									
$tc^3 = tc^2 \cap i$	1	0		0	$\overline{D}$	0	1	$\overline{D}$	$\overline{D}$
$tc^4 = tc^3 \cap h$	1	0	1	0	$\overline{D}$	0	1	$\overline{D}$	$\overline{D}$
$tc^5 = tc^4 \cap c$	$\varnothing$	0	1	0	$\overline{D}$	0	1	$\overline{D}$	$\overline{D}$

Table 2.5   D-Drive Along Lines 5–7–9 and Consistency

	1	2	3	4	5	6	7	8	9
D-drive									
tc^0	1	0			$\overline{D}$				
$tc^1 = tc^0 \cap f_d$	1	0		1	$\overline{D}$		D		
$tc^2 = tc^1 \cap j_d$	1	0		1	$\overline{D}$		D	1	D
Consistency									
$tc^3 = tc^2 \cap l$	1	0		1	$\overline{D}$	1	D	1	D
$tc^4 = tc^3 \cap g$	1	0	0	1	$\overline{D}$	1	D	1	D

37

2.5 with the consistency operations. It can be seen from Table 2.5 that the final test cube is tc^4. Thus, 100 is a test for detecting the fault 5 s-a-1.

The D-algorithm generates a test for every fault in a circuit, if such a test exists. It uses less computation time and less memory space, and hence it is more efficient than the Boolean difference method. It can also identify redundant faults by "proving" that no corresponding test exist.

2.2.4 PODEM (Path-Oriented Decision-Making)

PODEM is an enumeration algorithm in which all input patterns are examined as tests for a given fault [2.8]. The search for a test continues till the search space is exhausted or a test pattern is found. If no test pattern is found, the fault is considered to be untestable. PODEM uses logic schematic diagrams in a manner similar to the D-algorithm for deriving tests. The high-level description of PODEM is shown in Fig. 2.12. The functions of each box in the diagram are as explained below.

Box 1   Initially, all primary inputs (PIs) are at x; that is, they are unassigned. One of the PIs is assigned a 0 or 1, and the PI is recorded as an unflagged node in a decision tree. Thus, the process is similar to branch in the context of branch and bound algorithms.

Box 2   The value at the selected primary input is forward traced in conjunction with x's at the rest of the primary inputs, by using the five-valued logic 0, 1, x, $D, \overline{D}$.

Box 3   If the input pattern of Box 2 constitutes a test, then the test generation process is completed.

Box 4   The decision tree increases in depth; that is, one more primary input is assigned a 0 or 1 to check if it is possible to generate a test. Two possible situations may arise while evaluating Box 4:

1. The signal line (on which a stuck-at fault is assumed to be present) has the same logic value as the stuck-at value.
2. There is no signal path from an internal signal line to a primary output such that the line is at D or $\overline{D}$ and all the lines are at x.

In the first case, the fault remains masked in the presence of the assigned input values, whereas in the second case the input pattern cannot be a test, because D or $\overline{D}$ cannot be propagated to the output. Therefore, only when none of the foregoing situations occurs is a test possible with the current assignment of PIs.

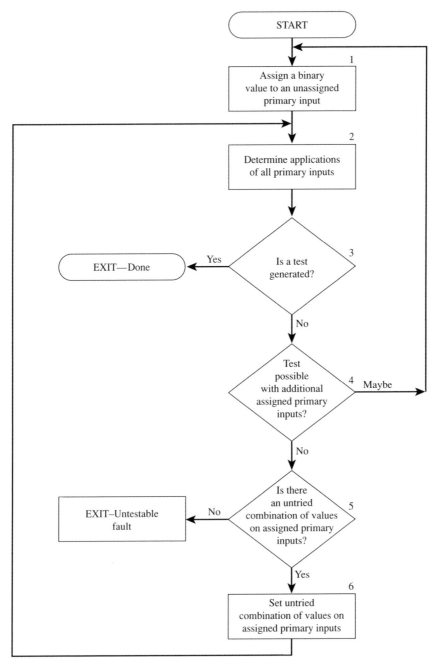

Figure 2.12 PODEM algorithm (from P. Goel, ''An implict enumeration algorithm to generate tests for combinational circuits,'' IEEE Trans. Comput., March 1981. Copyright © 1981 IEEE. Reprinted with permission).

39

Box 5   If all primary inputs have been assigned values and a test pattern is still not found, it is checked whether an untried combination values at the inputs might generate a test or not.

It is clear from the preceding discussion that the decision tree is an ordered list of nodes (Fig. 2.13) having the following features:

1. Each node identifies a current assignment of 0 or 1 to a primary input.
2. The ordering reflects the sequence in which the current assignments have been made.

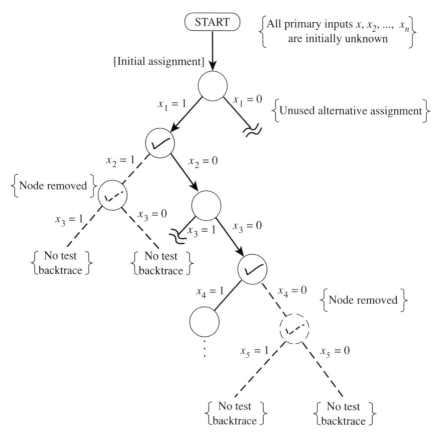

Figure 2.13   Decision tree in PODEM algorithm (from P. Goel, ''An implicit enumeration algorithm to generate tests for combinational circuits,'' IEEE Trans. Comput., March 1981. Copyright © 1981 IEEE. Reprinted with permission).

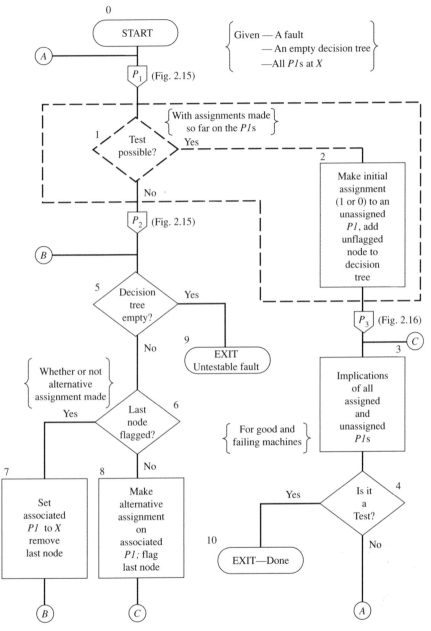

Figure 2.14 Flowchart of PODEM algorithm (from P. Goel, ''An implict enumeration algorithm to generate tests for combinational circuits,'' IEEE Trans. Comput., March 1981. Copyright © 1981 IEEE. Reprinted with permission).

A node is *flagged* (indicated by a check mark inside the node) if its initial assignment has been rejected and the alternative is being tried. When both assignment choices at a node are rejected, the associated node is removed and the predecessor nodes assignment is also rejected. The assignment at the most recently selected PI is rejected if no test can be formed with this assignment. The rejection of a PI assignment results in a *bounding* of the decision tree, because it avoids the enumeration of subsequent assignments to as yet unassigned PIs.

Figure 2.14 shows a more detailed description of the PODEM. The decision tree of Box 5 can be implemented as a stack. An initial PI assignment pushes an unflagged node onto the stack. The bounding of the decision tree is done by popping the stack until an unflagged node is on the top of the stack (Box 2). The assignment value to that node is complemented, and the node is flagged (Box 8). All nodes popped out of the stack are in effect removed from the decision tree, and then associated PIs are set to x (Box 7). A new combination of values on the PIs is obtained by the bounding process, and it is evaluated (Boxes 3 and 4) as a possible test pattern. The entire process is iteratively continued until a test pattern is found, or it is determined that a test is possible only with additional PI assignments (Box 2), or the decision tree is empty (Box 9). The decision tree becomes empty only if the fault under consideration is untestable. Because it is desirable to have least number of flagged nodes in a decision tree, the selection of a proper initial assignment will reduce the test generation time. PODEM uses a two-step process to choose a PI and its logic value assignment:

1. Determine an initial output. [An *objective* is defined by a logic value (0 or 1), referred to as *objective logic level*; the signal line on which the objective level is desired is known as *objective net*.]
2. Given the initial objective, choose a PI and its logic value that has a good likelihood of satisfying the initial objective.

Initial Objective

The flowchart of Fig. 2.15 shows the procedure to determine initial objectives. As mentioned previously, all signal lines in a circuit are initially unspecified. A stuck-at fault for which a test pattern is to be derived could be located either at the output or at an input of a gate, so that the output of the gate assumes (D or $\overline{D}$). As the algorithm continues, the gate under test (G.U.T.) may not have unspecified value at its output due to the implications of assignments at primary inputs (derived during the backtrace procedure).

As can be seen in Fig. 2.15, if the fault is assumed to be at the output of a

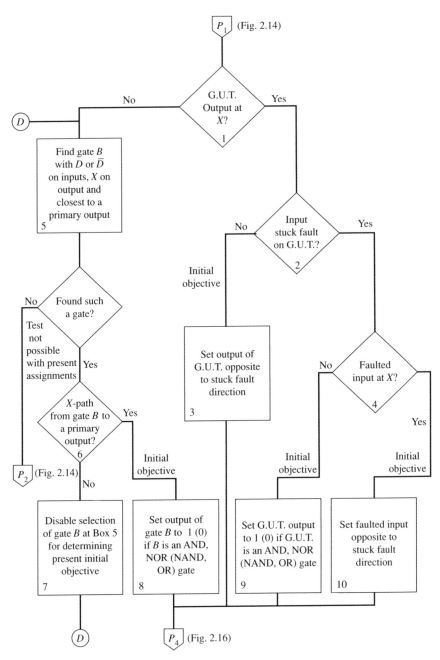

Figure 2.15 Determination of initial objective (from P. Goel, ''An implicit enumeration algorithm to generate tests for combinational circuits,'' IEEE Trans. Comput., March 1981. Copyright © 1981 IEEE. Reprinted with permission).

43

gate that is still unspecified, then the output is set to a value opposite to the assumed stuck-at value. On the other hand, if a fault is on input line of a gate, and the faulty input line is already specified, the output line has to be assigned 1 if the gate is AND (or NOR), and 0 if the gate is NAND (or OR). If the output of the gate under test is specified, that is, D ($\overline{D}$) not x, then the output is propagated via other gates toward a primary output. This can be done via a gate closest to a primary output. If such a gate is not found, the D or $\overline{D}$ cannot be propagated to the output, that is, a test pattern cannot be formed with the current assignment at the primary inputs. On the other hand, if a gate is available for propagating D or $\overline{D}$ to the output, it will still be necessary to have a path consisting of all xs (x-path) through which the D or $\overline{D}$ from the gate under test can be propagated to a primary output. If an x-path cannot be found, the gate should not be considered for determining the initial objective. However, if an x-path exists, the output of the gate is set to 1 if it is AND (or NOR), and 0 if it is NAND (or OR).

Backtrace

The procedure for obtaining a primary input assignment given an initial objective is shown in Fig. 2.16; this is known as *backtrace*. The backtrace procedure traces a signal path from an objective net with a given objective level (initial objective) backward to a primary input. During the backtracking, each gate is assigned logic values such that the desired initial objective at the objective net can be achieved. If the objective net is driven by an OR or a NAND gate, and the current objective level is 1 (0), the next objective net is the input line of the driving gate that is easiest (hardest) to control. Alternatively, if the objective net is driven by an AND or a NOR gate, and the current objective level is 0 (1), the input line of the driving gate that is easiest (hardest) to control is selected as the next objective net.

Let us demonstrate the application of the PODEM algorithm by deriving a test for detecting the fault α s-a-0 in the circuit of Fig. 2.17. The initial objective is to set the output of gate A to logic 1, that is, the objective logic level is 1 on net 5 (Box 3 in Fig. 2.15). By going through the backtrace procedure, it can be determined that the next objective net is 1 (or 2) and the objective logic level is 0. Because net 1 is fed by the primary input x_1, the current objective logic level, namely, logic 0, is assigned to the primary input x_1 as shown here:

1	2	3	4	5	6	7	8	9	10	11	12
0	X	X	X	X	X	X	X	X	X	X	X

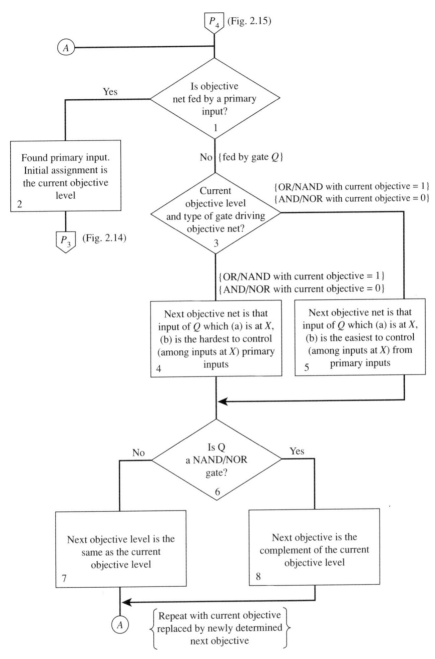

Figure 2.16 Backtrace procedure (from P. Goel, "An implicit enumeration algorithm to generate tests for combinational circuits," IEEE Trans. Comput., March 1981. Copyright © 1981 IEEE. Reprinted with permission).

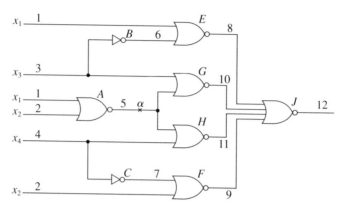

Figure 2.17   Circuit under test

Because $x_1 x_2 x_3 x_4 = 0XXX$ is not a test for the fault, a second pass through the algorithm results in the assignment of primary input x_2, which sets up D as the output of gate A:

1	2	3	4	5	6	7	8	9	10	11	12
0	0	X	X	D	X	X	X	X	X	X	X

Because the output of gate A, namely, net 5, is not X, it is necessary to find a gate with D as its input, X as its output, and closer to the primary output (Box 5, Fig. 2.15). Both gates G and H satisfy the requirements. The selection of gate G and the subsequent initial objective (Box 8, Fig. 2.15) result in the assignment of primary input x_3:

1	2	3	4	5	6	7	8	9	10	11	12
0	0	0	X	D	1	X	0	X	$\bar{D}$	X	X

$x_1 x_2 x_3 x_4 = 000X$ is not a test for the fault because the primary output is X. Gate J has D on input net 10 and Xs on input nets 9 and 11. The initial objective is to set the objective net 12 to logic 1. The selection of net 9 as the next objective results in the assignment of primary input x_4:

1	2	3	4	5	6	7	8	9	10	11	12
0	0	0	0	D	1	1	0	0	$\bar{D}$	$\bar{D}$	D

Thus, the test for the fault α s-a-0 is $x_1 x_2 x_3 x_4 = 0000$. The same test could be found for the fault by applying the D-algorithm; however, the D-algorithm requires substantial trial and error before the test is found. This is because of the variety of propagation paths and the attendant consistency operations that are required. For example, α s-a-0 has to be simultaneously propagated to the output via the paths *AGJ* and *AHJ*; propagation along either path individually will lead to inconsistency. This feature of the *D*-algorithm can lead to a waste of effort if a given fault is untestable. The PODEM is more efficient than the D-Algorithm in terms of computer time required to generate tests for combinational circuits.

2.2.5   FAN (Fanout-Oriented Test Generation)

The FAN algorithm is in principle similar to PODEM, but more efficient [2.9]. The efficiency is achieved by reducing the number of backtracks in the search tree. Unlike PODEM, where the backtracking is done along a single path, FAN uses the concept of multiple backtrace. Before we show how FAN deals with the test generation problem for stuck-at faults, several terms have to be defined. A *bound line* is a gate output that is part of reconvergent fan-out loop. A line that is not bound is considered to be *free*. A *headline* is a free line that drives a gate that is part of a reconvergent fan-out loop. In Fig. 2.18, for example, nodes *H*, *I*, and *J* are bound lines, *A* through *H* are free lines, and *G*, *H*, and *F* are headlines. Because by definition headlines are free lines, they can be considered as primary input lines and can always be assigned values arbitrarily. Thus, during the back-trace operation, if a headline is reached, the backtree stops; it is not necessary to reach a primary input to complete the backtrace.

FAN uses a technique called multiple backtrace to reduce the number of back-tracks that must be made during the search process. For example, in Fig. 2.19 if the objective is to set *H* at logic 1, PODEM would backtrace along one of the paths to the primary inputs. Suppose the backtrace is done via the path *H–E–C*, which will set *E* to 1. Because *E* is at 1, *C* will set to 0. However, a 0 at *C* sets *F* to 1, *G* to 0, and *H* to 0. Because this assignment fails to achieve the desired objective, the backtrace process is performed via another path, for example, *H–G–F–C*, and the desired goal can be achieved. Thus, in PODEM, several backtracks may be necessary before the requirement of setting up a particular

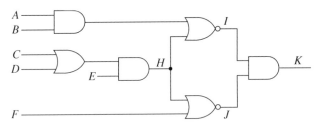

Figure 2.18   Example circuit

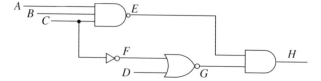

Figure 2.19 Multiple backtrace along H-E-C and H-G-F-C

logic value on a line is satisfied. FAN avoids this waste of computation time by backtracking along multiple paths to the fan-out point. For example, if multiple backtrace is done via both $H–E–C$ and $H–G–F–C$, the value at C can be set so that the value at H is justified.

We illustrate the application of the FAN algorithm by deriving a test for the fault Z s-a-0 in Fig. 2.20. First, the value D is assigned to the line Z and the value 1 to each of the inputs M and N. The initial objectives are to set M and N to 1. By the multiple backtrace, G and I are assigned 1 (note that instead of G and I, L could be assigned logic 1). Again, by the multiple backtrace, we have the final objectives $A = 1, B = 1$, and $E = 1, F = 1$. The assignment $A = 1, B = 1$ makes $J = 1, M = 1$, and the assignment $E = 1, F = 1$ makes $I = 1, N = 1$. Thus, the assignments $A = B = E = F = 1$ constitute a test for the fault Z s-a-0. It is easy to see that if the first multiple backtrace stopped at L, and the second multiple backtrace at H, the test for the fault will be $C = D = 1$.

2.2.6 DELAY FAULT DETECTION

A delay fault in a combinational logic circuit can be detected only by applying a sequence of two test patterns. The first pattern, known as an *initialization pattern,* sets up the initial condition in a circuit so that the fault slow-to-rise or slow-to-

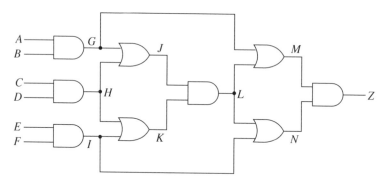

Figure 2.20 Circuit under test

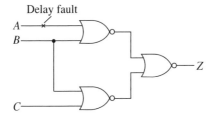

Figure 2.21   Circuit with a delay fault

fall signal at the input or output of a gate can affect an output of the circuit. The second pattern, known as a *transition* or *propagation pattern,* propagates the effect of the activated transition to a primary output of the circuit. To illustrate, let us consider a delay (slow-to-rise) fault at the input A of the circuit shown in Fig. 2.21. The test for slow-to-rise fault consists of the initialization pattern $ABC = 001$, followed by the transition pattern $ABC = 101$. Similarly, the two pattern tests for a slow-to-fall delay fault at input A will be $ABC = 101, 001$. Note that the slow-to-rise fault (slow-to-fall fault) corresponds to a transient stuck-at-0 (stuck-at-1) fault.

To identify the presence of a delay fault in a combinational circuit, the hardware model shown in Fig. 2.22 is frequently used in literature. The initialization pattern is first loaded into the input latches. After the circuit has stabilized, the transition pattern is clocked into the input latches by using $C1$. The output pattern of the circuit is next loaded into the output latches by setting the clock $C2$ at logic 1 for a period equal to or greater than the time required for the output pattern to be loaded into the latch and stabilize. The possible presence of a delay fault is confirmed if the output value is different from the expected value.

Delay tests can be classified into two groups: nonrobust and robust [2.10]. A delay fault is *nonrobust* if it can detect a fault in the path under consideration provided there are no delay faults along other paths. For example, the input vector pair (111, 101) can detect the slow-to-rise fault at e in Fig. 2.23(a) as long as the

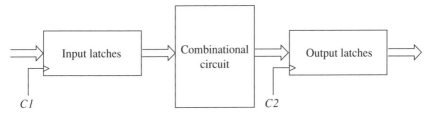

Figure 2.22   Hardware model for delay fault testing

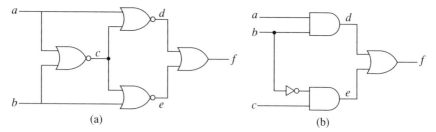

Figure 2.23　The illustration of (a) Nonrobust test; (b) robust test

path b–d–f does not have a delay fault. However, if there is a slow-to-fall fault at d, the output of the circuit will be correct for the input pair, thereby invalidating the test for the delay fault at e. Therefore, the test (111, 101) is nonrobust.

　　A delay test is considered to be *robust* if it detects the fault in a path independent of delay faults that may exist in other paths of the circuit. For example, let us assume a slow-to-fall delay fault at d in the path a–c–d–f of the circuit shown in Fig. 2.23(b). The input vector pair (01, 11) constitutes a robust test for the delay fault because the output of any gate on the other paths does not change when the second vector of the input pair is applied to the circuit. Thus, any possible delay fault in these paths will not affect the circuit output. Robust tests do not exist for many paths in large circuits [2,11,2.12]. Significant research in recent years has concentrated on the design of circuits that are fully testable for all path delay faults using robust tests [2.13–2.16].

2.3　Detection of Multiple Faults in Combinational Logic Circuits

One of the assumptions normally made in test generation schemes is that only a single fault is present in the circuit under test. This assumption is valid only if the circuit is frequently tested, when the probability of more than one fault occurring is small. However, this is not true when a newly manufactured circuit is tested for the first time. Multiple-fault assumption is also more realistic in the VLSI environment, where faults occurring during manufacture frequently affect several parts of a circuit. Some statistical studies have shown that multiple faults, composed of at least six single faults, must be tested in a chip to establish its reliability [2.17].

　　Designing multiple-fault detection tests for a logic network is difficult because of the extremely large number of faults that have to be considered. In a circuit having k lines there are $2k$ possible single faults, but a total of $3^k - 1$ multiple faults [2.18]. Hence, test generation for all possible multiple faults is impractical even for small networks.

One approach that reduces the number of faults that need be tested in a network is *fault collapsing* which uses the concept of *equivalent faults* [2.19,2.20]. For example, an x-input logic gate can have $2x + 2$ possible faults; however, for certain input faults, a gate output would be forced into a state that is indistinguishable from one of the s-a-0/s-a-1 output faults. Thus, for an AND gate any input s-a-0 fault is indistinguishable from the output s-a-0 fault, and for an OR gate any input s-a-1 fault is indistinguishable for the output s-a-1 fault. Such faults are said to be equivalent. For a NAND (NOR) gate, the set of input s-a-0 (s-a-1) faults and the set of output faults s-a-1 (s-a-0) are equivalent. Thus, an x-input gate has to be tested for $x + 2$ logically distinct faults.

A systematic approach that reduces the number of faults that have to be considered in test generation is the process of *fault folding* [2.21]. The central idea behind the process is to form fault equivalence classes for a given circuit by folding faults toward the primary inputs. For nonreconvergent fan-out circuits, the folding operation produces a set of faults on primary inputs, and this set test covers all faults in the circuit. For reconvergent fan-out circuits, the set of faults at the primary inputs, fan-out origins, and fan-out branches test cover all faults in the circuit.

Another approach that results in a significant reduction in the number of faults to be tested uses the concept of *prime faults* [2.22]. The set of prime faults for a network can be generated by the following procedure:

1. Assign a fault to every gate input line if that is a primary input line or a fan-out branch line. The fault is s-a-1 for AND/NAND gate inputs, and s-a-0 for OR/NOR gate inputs. Treat an inverter as a single input NAND/NOR gate if its output is a primary output; otherwise, no fault value should be assigned to an inverter input line.

2. Identify every gate that has faults assigned to all its input lines as a *prime gate*. Assign a fault to the output line of every prime gate that does not fan out. The fault is s-a-0 for AND/NOR gate outputs, and s-a-1 for OR/NAND gate outputs.

The number of prime faults in a network is significantly fewer than the number of single faults, because many single faults can be represented by equivalent multiple prime faults.

In general, test sets derived under the single-fault assumption can detect a large number of multiple faults. However, there is no guarantee that a multiple

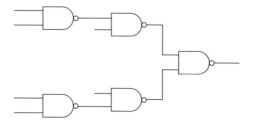

Figure 2.24   Multilevel fan-out-free network

fault will be detected by a single-fault-detecting test set, although all the single-fault components of the multiple fault may be individually detected; this is due to masking relations between faults [2.23].

Schertz *et al.* [2.24] have shown that for a *restricted fan-out-free* network, every single-fault detection test set is also a multiple-fault detection test set. Any network that does not contain the network of Fig. 2.24 (or its equivalent) as a subnetwork is called a *restricted fan-out-free* network. It has been proved that the network of Fig. 2.24 is the smallest fan-out-free network that can contain multiple faults that are not detected by every single-fault detection test set [2.18].

Bossen and Hong [2.25] have shown that any multiple fault in a network can be represented by an equivalent fault, the components of which are faults on certain *checkpoints* in the network. Checkpoints are associated with each fan-out point and each primary input.

Agarwal and Fung [2.26] have shown that the commonly used hypothesis about testing multiple faults by single-fault tests is not correct for reconvergent fan-out circuits. They suggested that the only way to test for multiple faults in VLSI chips is to design them so that they will be easily testable for multiple faults.

References

2.1   Levendel, Y. H., and P. R. Menon, "Fault simulation methods—extension and comparison," *Bell Syst. Tech. Jour.*, 2235–2258 (November 1981).

2.2   Goel, P. "Test generation cost analysis and projections," *Proc. 17th Design Automation Conf.*, 77–84 (1980).

2.3   Armstrong, D. B., "On finding a nearly minimal set of fault detection tests for combinational logic nets," *IEEE Trans. Electron. Comput.*, 66–73 (February 1966).

2.4 Schneider, R. R., "On the necessity to examine *D*-chains in diagnostic test generation," *IBM Jour. of Res. and Develop.*, 114 (January 1967).

2.5 Sellers, F. F., M. Y. Hsiao, and C. L. Bearnson, "Analyzing errors with the Boolean difference," *IEEE Trans. Comput.*, 676–683 (July 1968).

2.6 Chiang, A. C., I. S. Reed, and A. V. Banes. "Path sensitization, partial Boolean difference and automated fault diagnosis," *IEEE Trans. Comput.*, 189–195 (February 1972).

2.7 Roth, J. P., "Diagnosis of automata failures: A calculus and a method," *IEEE Trans. Comput.*, 278–291 (July 1966).

2.8 Goel, P., "An implicit enumeration algorithm to generate tests for combinational logic circuits," *IEEE Trans. Comput.*, 215–222 (Marcy 1981).

2.9 Fujiwara, H., and T. Shimono, "On the acceleration of test generation algorithms," *IEEE Trans. Comput.*, 1137–1144 (December 1983).

2.10 Park, E., and M. Mercer, "Robust and nonrobust tests for path delay faults in a combinational circuit," *Proc. Intl. Test Conf.*, pp. 1027–1034 (1987).

2.11 Reddy, S. M., C. Lin and S. Patil, "An automatic test pattern generator for the detection of path delay faults," *Proc. IEEE Intl. Conf. on CAD*, pp. 284–287 (November 1987).

2.12 Schulz, M. H., K. Fuchs and F. Fink, "Advanced automatic test pattern generation techniques for path delay faults," *Proc. 19th IEEE Intl. Fault-Tolerant Computing Symp.*, pp. 44–51 (June 1989).

2.13 Kundu, S., and S. M. Reddy, "On the design of robust testable CMOS combinational logic circuits," *Proc. 18th Intl. Fault-Tolerant Computing Symp.*, pp. 220–225 (June 1988).

2.14 Roy, K., J. A. Abraham, K. De, and S. Lusky, "Synthesis of delay fault testable combinational logic," *Proc. IEEE Intl. Conference on CAD*, pp. 418–421 (November 1989).

2.15 Pramanick, A. K., S. M. Reddy, and S. Sengubta, "Synthesis of combinational logic circuits for path delay fault testability," *Proc. Intl. Symp. on Circuits and Systems*, pp. 3105–3108 (May 1990).

2.16 Pramanick, A. K., and S. M. Reddy, "On the design of path delay fault testable combinational circuits," *Proc. IEEE Intl. Fault-Tolerant Computing Symp.*, pp. 374–381 (June 1990).

2.17 Goldstein, L. H., "A probabilistic analysis of multiple faults in LSI circuits, *IEEE Computer Society Repository*, R77–304 (1977).

2.18 Hayes, J. P., "A NAND model for fault diagnosis in combinatorial logic circuits," *IEEE Trans. Comput.*, 1496–1506 (December 1971).

2.19 McCluskey, E. J., and F. W. Clegg, "Fault equivalence in combinational logic networks," *IEEE Trans. Comput.*, 1286–1293 (November 1971).

2.20 Schertz, D. R., and G. A. Metze, "A new representation for faults in combinational digital circuits," *IEEE Trans. Comput.*, 858–866 (August 1972).

2.21 Klin To, "Fault-folding for irredundant and redundant combinational net-works," *IEEE Trans. Comput.*, 1008–1015 (November 1973).

2.22 Cha, C. W., "Multiple fault diagnosis in combinational networks," *Proc. 16th Design Automation. Conf.*, 149–155 (1979).

2.23 Dias, F.J.O., "Fault masking in combinational logic circuits," *IEEE Trans. Comput.*, 476–482 (May 1975).

2.24 Schertz, D. R., and G. Metze, "On the design of multiple fault diagnosable networks," *IEEE Trans. Comput.*, 1361–1364 (November 1971).

2.25 Bossen, D. C., and S. J. Hong, "Cause–effect analysis for multiple fault detection in combinational networks," *IEEE Trans. Comput.*, 1252–1257 (November 1971).

2.26 Agarwal, V. K., and A. S. Fung, "Multiple fault testing of logic circuits by single fault test sets," *IEEE Trans. Comput.*, 854–855 (November 1981).

Chapter 3 | Testable Combinational Logic Circuit Design

A logic circuit is considered to be testable if it is easy to generate a set of test patterns to achieve high fault coverage in the circuit. In recent years, a number of design techniques have been proposed for the realization of testable combinational logic circuits. These techniques consider mainly *unstructured*, that is, gate-level, implementation of combinational circuits. At the VLSI level, PLAs (Programmable Logic Arrays) are often used for the implementation of combinational logic functions. Although PLAs in general need more area than the gate or standard cell implementations, they have the advantage of memory-like regular structure. In this chapter, we discuss several techniques for designing testable combinational logic at the gate level. Also, several techniques for testable PLA design are discussed.

3.1 The Reed–Muller Expansion Technique

This technique can be used to realize any arbitrary n-variable Boolean function using AND and EX–OR gates only. The circuit so designed has the following properties:

1. If the primary input leads are fault-free, then at most $n + 4$ tests are required to detect all single stuck-at faults in the circuit.
2. If there are faults on the primary input leads as well, then the number of tests required is $(n + 4) + 2n_e$, where n_e is the number of input variables that appear an even number of times in the product terms of the Reed–Muller expansion. However, by adding an extra AND gate with its output

being made observable, the additional $2n_e$ tests can be removed. The input to the AND gate are those inputs appearing an even number of times in the Reed–Muller product terms.

Any combinational function of n variables can be described by a Reed–Muller expansion of the form

$$f(x_1, x_2, \ldots, x_n) = C_0 \oplus C_1 \dot{x}_1 \oplus C_2 \dot{x}_2 \oplus \cdots \oplus C_n \dot{x}_n \oplus C_{n+1} \dot{x}_1 \dot{x}_2$$
$$\oplus C_{n+2} \dot{x}_1 \dot{x}_3 \oplus \cdots \oplus C_{2^n-1} \dot{x}_1 \dot{x}_2 \ldots \dot{x}_n,$$

where $\dot{x}_i$ is either x_i or $\bar{x}_i$ but not both together, C_i is a binary constant 0 or 1 and $\oplus$ is the modulo-2 sum (exclusive-OR operation). If all $\dot{x}_i = x_i$, this special case is known as the *complement-free ring sum* expansion of the Boolean function.

For a three-variable function, the Reed–Muller expansion is

$$f(W, X, Y) = C_0 \oplus C_1 W \oplus C_2 X \oplus C_3 Y$$
$$\oplus C_4 WX \oplus C_5 WY \oplus C_6 XY \oplus C_7 WXY.$$

The constants C_i for a Reed–Muller expansion may be computed by using the following properties of the EX–OR operation:

$$\bar{A} = 1 \oplus A,$$

$$A + B = A \oplus B \oplus AB.$$

To illustrate, let us consider the Boolean function

$$f(W, X, Y) = WX + \overline{W}Y + \overline{X}\overline{Y}.$$

This can be represented as

$$f(W, X, Y) = WX \oplus (1 \oplus W)Y \oplus (1 \oplus X)(1 \oplus Y).$$

Thus, the Reed–Muller expansion of the function is:

$$f(W, X, Y) = 1 \oplus X \oplus WX \oplus WY \oplus XY.$$

A direct implementation of the function is shown in Fig. 3.1.

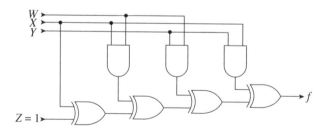

Figure 3.1   Reed–Muller circuit for $f = WX + WY + XY$

It has been shown that to detect a single faulty gate in a cascade of exclusive-OR gates it is sufficient to apply a set of test inputs that will exercise each exclusive-OR gate for all possible input combinations. Such a test set for the circuit of Fig. 3.1 is given by

$$
\begin{array}{cccc}
Z & W & X & Y
\end{array}
$$
$$
T_1 = \left\{\begin{array}{cccc}
0 & 0 & 0 & 0 \\
0 & 1 & 1 & 1 \\
1 & 0 & 0 & 0 \\
1 & 1 & 1 & 1
\end{array}\right\}
$$

The structure of the test is always the same, independent of the number of input variables, and constitutes four tests only; for example, a five-variable circuit would have the test set

$$
\begin{array}{cccccc}
Z & U & V & W & X & Y
\end{array}
$$
$$
\left\{\begin{array}{cccccc}
0 & 0 & 0 & 0 & 0 & 0 \\
0 & 1 & 1 & 1 & 1 & 1 \\
1 & 0 & 0 & 0 & 0 & 0 \\
1 & 1 & 1 & 1 & 1 & 1
\end{array}\right\}
$$

In addition the test set T_1 will also detect

1. Any s-a-0 fault on the input or output of an AND gate (tests 0111, 1111)
2. Any s-a-1 fault on the output of an AND gate (tests 0000, 1000)

However, an s-a-1 fault on the AND gate inputs must be detected separately using the test set

$$
\begin{array}{cccc}
Z & W & X & Y
\end{array}
$$
$$
T_2 = \left\{\begin{array}{cccc}
- & 0 & 1 & 1 \\
- & 1 & 0 & 1 \\
- & 1 & 1 & 0
\end{array}\right\}
$$

where "$-$" is a don't care condition. Thus, for an n-variable function, T_2 will contain n tests and the full test set will now consist of $T = T_1 + T_2$ and contain $n + 4$ tests.

It has already been mentioned that if the faults on the primary input lines are also considered, the number of tests increases by $2n_e$. In our example, $n_e = 2$, because the variables W and Y occur twice. However, by incorporating an AND gate in the circuit of Fig. 3.1 such that the inputs to the AND gate are W and Y,

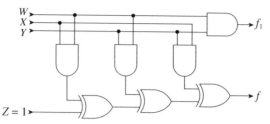

Figure 3.2   Modified Reed–Muller circuit for $f = \overline{W}X + WY + \overline{X}\overline{Y}$

the original $n + 4$ tests can be used to detect both primary input and gate faults. The modified circuit is shown in Fig. 3.2; the output f_1 is observable.

3.2   Three-Level OR–AND–OR Design

One of the major drawbacks of the design technique based on Reed–Muller expansion is the increased number of logic levels in a circuit. An alternative design method that results in AND–OR networks of at most three levels, and in which single s-a-1 or s-a-0 faults are locatable within certain indistinguishable classes, will now be discussed [3.1]. The method is applicable only to positive unate logic functions. A logic function is *unate* if it can be represented as a sum-of-products or product-of-sums expression in which each variable appears either in a complemented form or in an uncomplemented form but not in both. A positive unate function is one in which all variables are uncomplemented. The design process commences with the design of nonredundant three-level OR–AND–OR *prime trees* [3.2]. A *tree network* is a logic circuit in which the output of any gate is not fed to more than one gate, that is, no fan-out is allowed except on primary inputs. A *restricted tree network* is a tree network composed of AND, OR, and inverter gates with the restriction that inverters are fed only from external inputs. A *restricted prime tree* is a restricted tree network satisfying the following assumptions, where f represents the output of the network:

1. If the output gate is an OR gate and if

$$f = T_1 + T_2 + \cdots + T_p,$$

     where T_i is a product term, then T_i is a prime implicant of f, $1 \leqslant i \leqslant p$.

2. If the output gate is an AND gate and if

$$f = U_1 \cdot U_2 \cdots U_q,$$

     where U_i is a sum term, then U_i is a prime implicant of f, $1 \leqslant i \leqslant q$.

A *prime tree* is a tree network containing AND, OR, NAND, NOR, and inverter gates, which is either a restricted prime tree or has a test-equivalent network that is a restricted prime tree. A way to design prime tree networks is to start with a sum of prime implicants or a product of prime implicants for the given function and then use the following factoring techniques:

$$ab + ac = a(b + c),$$

$$(a + b)(a + c) = a + bc.$$

A prime tree is nonredundant if no lead in the prime tree can be connected to a logical constant, 1 or 0, without changing the functional value of the output. In three-level tree networks, the two-level networks the outputs of which are connected to the third-level gates are called *subtrees*.

An OR–AND–OR tree can be formed by grouping the prime implicants according to the number of literals present in them, and then determining possible factorization within these groups. As an example, let us design an OR–AND–OR prime tree for the function

$$f(A, B, C, D, E, F) = BEF + BCF + ACF + BDE + ACDE + ABCD.$$

After factorization,

$$f(A, B, C, D, E, F) = BE(D + F) + CF(A + B) + ACD(B + E).$$

The logic implementation of the function is shown in Fig. 3.3.

The design procedure for completely fault-locatable networks for unate functions is as follows:

1. Design a nonredundant three-level OR–AND–OR prime tree.

2. Check if two subtrees can be replaced by a single subtree. Two subtrees can be replaced by a single subtree if, for two resolvable stuck-at-1 faults, the same prime implicants appear in the sum-of-product expressions for both the subtrees.

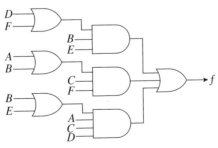

Figure 3.3   A prime tree for the function $f(A, B, C, D, E, F)$

3. Check if the tree network can be modified. In a nonredundant OR–AND–OR prime tree realization, if a stuck-at-1 fault at a primary input is indistinguishable from a stuck-at-1 fault at some other lead (not another primary input or the output) and this pair of faults is resolvable, then another prime tree can be obtained in which the primary input and the other lead are inputs to the same first-level OR gate.

3.3   Automatic Synthesis of Testable Logic

The objective of automatic logic synthesis is to generate a gate-level netlist from a design description at the register-transfer level (RTL). The synthesis process consists of two distinct phases: *translation* and *optimization*. During the translation process, a set of Boolean expressions is derived from the RTL description. These expressions are converted into a netlist form using two broad classes of techniques: *minimization* and *restructuring*. Ideally, the circuit implementation of the netlist should be nonredundant [3.3]. Two-level logic minimization is extremely important in Programmable Logic Array (PLA)–based combinational logic implementation, because this results in the minimum number of product terms, that is, rows in the PLA. In multilevel logic minimization, individual sum-of-products Boolean expressions are transformed into factored form that produces an equivalent logic representation with more levels. Restructuring reduces the number of levels in an expression by deriving a factored form representation of the expression. For example, the expression $f(a,b,c,d) = ac + bc + \overline{a}b\overline{c} + \overline{a}\,\overline{d}c + c\overline{d}$ can be represented as $(\overline{a} + c)(b + \overline{d}) + ac$. The implementations of the expressions are shown in Figs. 3.4(a) and (b), respectively.

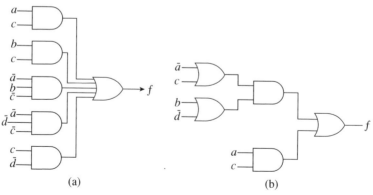

(a)   (b)

Figure 3.4   Implementations of (a) a Boolean expression and (b) its factored form

A factored form representation of a Boolean expression is obtained by using either *algebraic* or *Boolean* division. Let us first define certain terms that are necessary to explain the division process. A *literal* is a variable or its complement. A *cube* is a set of literals such that $x \in C$ implies $\bar{x} \notin C$. A cube represents the product of its literals. For example, three literals a, b, and c can form a cube abc. An *expression* is the sum of its cubes. For example, $\bar{b} + \bar{b}c$ is an expression. This expression is *redundant* because $\bar{b}$ *covers*, or contains, cube $\bar{b}c$. The *support* of an expression f is the set of variables on which f depends. For example, in the expression

$$f = a\bar{c} + bd + \bar{a}\bar{b} + \bar{b}c,$$

the support of f is $\{a,b,c,d\}$.

The product of two expressions f and g is an *algebraic product* if it contains no variables that are common to f and g, that is, expressions f and g have disjoint support. Alternatively, if the product of f and g have common support, it is known as a *Boolean product*.

The division of an expression f by another expression g can be represented as

$$f = gh + r,$$

where h and r are expressions; h is the *quotient* and r is the *remainder*. In a Boolean division, gh is a Boolean product, whereas in an algebraic division gh is an algebraic product.

The algebraic division of an expression f by another expression g with n terms can be performed by using the following steps [3.4]:

1. Divide f by each cube of g to generate n partial quotients ($h_1, h_2, \ldots, h_n$).

2. Select cubes that are common to all partial quotients to form the quotient h.

3. Place cubes that have not been used to form h in the remainder r.

Let us illustrate this procedure by dividing $f = bc + ace + de$ by $g = b + ae$.

1. Partial quotients $h_1 = c$, $h_2 = c$

2. Quotient $h = c$

3. Remainder $r = de$

Thus, $f = c(b + ae) + de$.

Algebraic divisors can contain a single cube or multiple cubes. An algebraic cube containing a single cube is called a *single cube divisor*, and one containing more than one cube is called a *multiple cube divisor*. In the preceding example, c is the single cube divisor, and $(b + ae)$ is a multiple cube divisor.

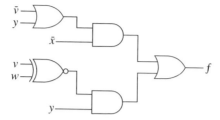

Figure 3.5   A multifault testable circuit

Hachtel [3.5] has proposed a synthesis technique that generates multiple-fault-testable combinational circuits. The technique consists of the following steps:

1. Convert the two-level representation of a combinational circuit into a *prime* and *irredundant* form. A circuit is prime and irredundant if no gate (an implicant) or its input (a literal) can be removed without causing the circuit to be functionally different.

2. Use *algebraic factoring* to derive the multilevel implementation of the circuit.

To illustrate, let us consider the four-variable function

$$f(v, w, x, y) = \bar{v}w\bar{x}y + \bar{v}x\bar{y} + \bar{w}\bar{x}y + \bar{v}\bar{w}y + vwy.$$

The minimized form of the expression is

$$f(v, w, x, y) = \bar{v}\bar{x} + \bar{v}\bar{w}y + \bar{x}y + vwy.$$

The expression can be factorized algebraically into

$$f(v, w, x, y) = \bar{x}(\bar{v} + y) + (\bar{v}\bar{w} + vw)y.$$

Figure 3.5 shows the implementation of the circuit.

This synthesis technique is also applicable to multioutput combinational circuits, provided each function corresponding to an output is made prime and irredundant. A multiple fault in a circuit has an equivalent multiple fault in an

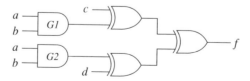

Figure 3.6   A circuit with redundant gates

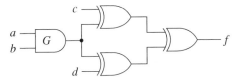

Figure 3.7   Multiple-fault testable implementation of the circuit of Fig. 3.6

algebraically factored version of the circuit. Since an algebraically factored version of a circuit results in a multilevel, area-efficient implementation of the circuit, this technique can be useful in designing multifault-testable circuit.

Logic optimization does not always guarantee complete testability. A circuit with duplicated gates may be completely single-fault-testable but is not so if the duplicated gates are replaced by a single gate [3.3]. To illustrate, let us consider the circuit (adapted from Ref. 3.3) shown in Fig. 3.6. Although either gate *G1* or *G2* can be removed from the circuit without affecting the circuit function, that is, the circuit is redundant, all single faults in the circuit are detectable. On the other hand, if *G1* and *G2* are replaced by a single gate *G* as shown in Fig. 3.7, a stuck-at-0 or a stuck-at-1 fault at the output of gate *G* is not detectable. Thus, the removal of redundancy has resulted in diminished testability for single faults in this circuit. However, a unidirectional multiple fault in Fig. 3.6, for example, the output of gate *G1* and *G2* stuck-at-0 or stuck-at-1, will not be detected. Thus, in Fig. 3.7 the multiple-fault testability is enhanced at the expense of reduced single-fault testability.

3.4 Testable Design of Multilevel Combinational Circuits

The starting point of the multilevel implementation of a function is the minimized two-level representation of the function. Rajski and Vasudevamurthy [3.6] have proposed a method for transforming two-level circuits into multilevel forms using only double cube divisors and single cube divisors with two literals, both in normal and complement form. The advantage of this transformation method is that the resulting multilevel circuit can be tested for all single stuck-at faults by the test set derived for the two-level circuit. The transformation method can be grouped into three categories:

1. Extraction of a single cube
2. Extraction of a double cube
3. Simultaneous extraction of a double cube and its complement

In each case, a set of tests that detects all single stuck-at faults in an original circuit will also detect all such faults in the transformed circuit. Let us consider each case separately:

3.4.1 SINGLE CUBE EXTRACTION

This involves identifying cubes that are common to two or more cubes in a sum-of-products expression. For example, let us consider the expression

$$f(a,b,c,d) = abc + bc\bar{d} + \bar{b}\bar{c}d + \bar{a}c.$$

Because cube bc is common to two product terms, the expression can be rewritten as

$$f(a,b,c,d) = (a + \bar{d})bc + \bar{b}\bar{c}d + \bar{a}c.$$

Figures 3.8(a) and (b) show the original and the transformed circuit, respectively.

3.4.2 DOUBLE CUBE DIVISOR

As the name indicates, a double cube divisor has exactly two cubes. The double cube divisors of an expression must be *cube-free*, that is, the divisors can be evenly divided only by 1. For example, $\bar{a}b + c$ is a double cube divisor because it has two cubes, and it is cube-free. On the other hand, $\bar{a}b + \bar{a}c$ is not cube-free (because it can be evenly divided by $\bar{a}$); thus, in spite of having two cubes, it is not a double cube divisor.

A double cube divisor is represented by $D_{x,y,z}$, where x is the number of literals in the first cube, y is the number of literals in the second cube, and z is the number

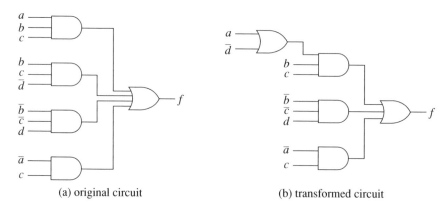

(a) original circuit   (b) transformed circuit

Figure 3.8   Single cube extraction

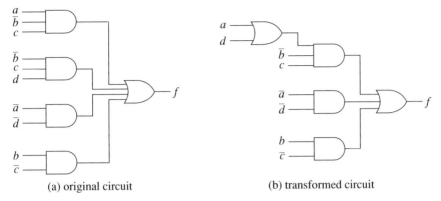

(a) original circuit   (b) transformed circuit

Figure 3.9   Double cube extraction

of support variables. For example, the double cube divisor $b + \bar{a}\bar{c}$ can be represented as $D_{1,2,3}$. Note that $D_{1,2,3}$ actually represents a subset of double cube divisors. Thus, the double cube divisor $b + ac$ also belongs to the subset $D_{1,2,3}$.

The extraction of a double cube divisor in a sum-of-products expression results in a circuit that can be tested for all stuck-at faults by the test set derived for the original circuit. To illustrate, let us consider the following expression:

$$f(a,b,c,d) = a\bar{b}c + \bar{b}cd + \bar{a}\bar{d} + b\bar{c}.$$

The extraction of the double cube divisor, $(a + d)$, results in

$$f(a,b,c,d) = (a + d)\bar{b}c + \bar{a}\bar{d} + b\bar{c}.$$

The implementation of the original and the transformed expressions are shown in Figs. 3.9(a) and (b), respectively.

3.4.3 EXTRACTION OF A DOUBLE CUBE DIVISOR
AND ITS COMPLEMENT

If an expression can be divided by a double cube divisor belonging to the subset $D_{1,1,2}$, and also by the complement of the divisor, then all single stuck-at faults in the circuit corresponding to the transformed expression can be detected by the test set derived for the original circuit. Note that the complement of a double cube divisor belonging to $D_{1,1,2}$ is a single cube divisor having two literals; this is because the double cube has a support of 2. If the support variables are x and y, the double cube divisors must be one of the following:

$$\bar{x} + \bar{y}, \qquad \bar{x} + y, \qquad x + y, \qquad x + \bar{y}.$$

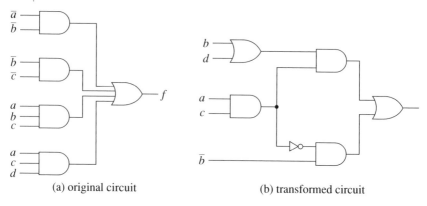

(a) original circuit   (b) transformed circuit

Figure 3.10   Extraction of a double cube and its complement

The complements of these double cube divisors are $xy, x\bar{y}, \overline{xy}$, and $\bar{x}y$, respectively, which are single cubes with two literals. Let us consider the following sum-of-products expression:

$$f(a,b,c,d) = \overline{a}\overline{b} + \overline{b}\overline{c} + abc + acd.$$

This can be transformed into

$$f(a,b,c,d) = (\overline{a} + \overline{c})\overline{b} + (b + d)ac$$

complement

Figures 3.10(a) and (b) show the original and the transformed circuit, respectively. The test patterns needed to completely test the circuit of Fig. 3.10(a) for all single

Table 3.1   **Test Set for the Circuit of Fig. 3.10(a)**

a	b	c	d	f
0	1	0	1	0
1	1	1	1	1
0	0	1	1	1
0	1	1	1	0
1	0	1	1	1
1	0	0	0	1
1	0	1	0	0
1	1	1	0	1
1	1	0	1	0

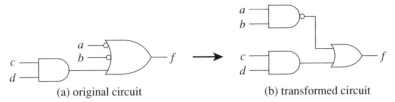

(a) original circuit   (b) transformed circuit

Figure 3.11   Fan-out-free transformation

stuck-at faults are shown in Table 3.1. The first eight tests of this test set also detect all single stuck-at faults in the transformed circuit of Fig. 3.10(b).

Batek and Hayes [3.7] have also presented an approach for transforming two-level combinational logic circuits to multilevel implementations that can be tested by a test set derived for the two-level circuits. The basic set of transformations for single-output circuits includes *fan-out-free transformation*, *extraction*, and *DeMorgan transformation*. A fan-out-free transformation converts a fan-out-free circuit into another fan-out-free circuit that is functionally equivalent to the original circuit. Figure 3.11 shows the fan-out-free transformation of the function $f(a,b,c,d) = \bar{a} + \bar{b} + cd$.

The extraction transformation converts an AND/OR structure to a simpler OR/AND version. Figure 3.12 illustrates the transformation of the AND–OR implementation of the function $f(a,b,c,d) = ab\bar{d} + bc\bar{d} + ac\bar{d}$ to the OR–AND implementation. Note that the combination of extraction and fan-out-free transformation is identical to the single cube and double cube extraction operations described previously.

The DeMorgan transformation transforms an AND–OR structure to an OR–AND structure. The OR–AND structure has an additional inverter at the output, and the number and the locations of the input inverters are also changed. Figure 3.13 illustrates such a transformation.

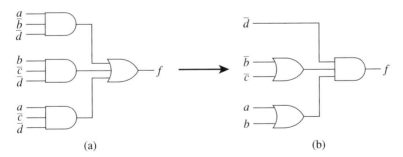

(a)   (b)

Figure 3.12   Extraction transformation

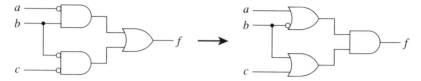

Figure 3.13   DeMorgan transformation

The foregoing transformations can be iteratively applied to parts of a circuit (or the corresponding expression), to generate an alternative version that can be tested by the test set derived for the original circuit. To illustrate let us consider the Boolean expression

$$f(a,b,c,d) = \overline{a}\,\overline{b} + \overline{b}\,\overline{d} + abd + acd + \overline{a}\,\overline{d}. \qquad (3.1)$$

that is implemented as shown in Fig. 3.14(a). The DeMorgan transformation results in the following modification of the expression (3.1):

$$f(a,b,c,d) = \overline{(b + ad)} + abd + acd + \overline{a}\,\overline{d}. \qquad (3.2)$$

The resulting circuit is shown in Fig. 3.14(b). Next, the extraction transformation is used to transform expression (3.2):

$$f(a,b,c,d) = \overline{(b + ad)} + ad(b + c) + \overline{a}\,\overline{d}. \qquad (3.3)$$

The circuit corresponding to expression (3.3) is shown in Fig. 3.14(c). Finally, the DeMorgan transformation can be used to modify the expression (3.3):

$$f(a,b,c,d) = \overline{(b + ad)} + ad(b + c) + \overline{(a + d)}. \qquad (3.4)$$

The resulting circuit is shown in Fig. 3.14(d). The complete test set for all detectable single stuck-at faults in the circuit of Fig. 3.14(a) is

a	b	c	d	f
0	1	0	1	0
0	0	1	1	1
0	1	1	0	1
1	0	1	1	0
1	0	0	0	1
1	1	1	0	1
1	1	0	0	1

The same test set also detects all detectable single stuck-at faults in the circuits of Figs. 3.14(b), (c), and (d). Thus, each modified version of the original circuit obtained by using the foregoing transformations preserves the original test set.

One additional transformation, *time-disjoint resubstitution*, may be used to

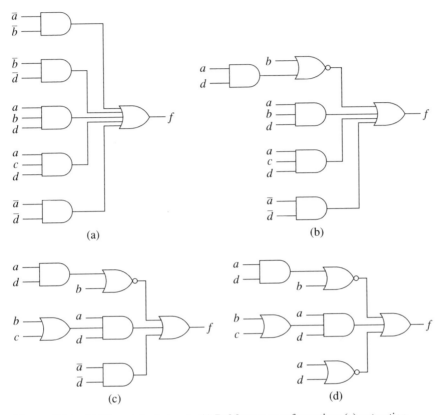

Figure 3.14   (a) The original circuit; (b) DeMorgan transformation; (c) extraction transformation; (d) DeMorgan transformation

replace separate copies of a subcircuit that does not fan out. The separate copies are replaced by a single copy the output of which is time-multiplexed to appropriate places by using a specific set of variables. Let us illustrate this transformation for the Boolean expressions of a four-input, two-output circuit:

$$f_1(a,b,c,d) = \bar{a}\bar{b}c + \bar{a}c\bar{d} + a\bar{b}\bar{c} + ac\bar{d},$$

$$f_2(a,b,c,d) = \bar{a}b\bar{c}d + abcd.$$

Note that output f_1 uses the subcircuit $(\bar{b} + \bar{d})$, and output f_2 uses the subcircuit bd, that is the complement of $(\bar{b} + \bar{d})$. Thus, the expressions can be rewritten as

$$f_1(a,b,c,d) = (\bar{a}c + a\bar{c})(\bar{b} + \bar{d}),$$

$$f_2(a,b,c,d) = (ac + \bar{a}\bar{c})\overline{(\bar{b} + \bar{d})}.$$

The implementation of the expressions is shown in Fig. 3.15.

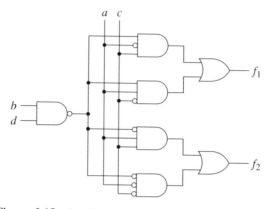

Figure 3.15   Application of resubstitution transformation

3.5   Synthesis of Random Pattern Testable Combinational Circuits

It is generally accepted that a minimized circuit is much easier to test than its nonminimized counterpart. This is because the minimization process simplifies a circuit by eliminating redundancy, which may introduce undetectable faults. However, redundancy is not necessarily the source of undetectable faults in a circuit. Let us first discuss the concept and the types of redundancy [3.8]. A gate-level circuit can be represented by a directed graph in which the vertices correspond to the gates, and the edges specify the interconnections of the gates. If at least one edge in the graph can be replaced by a constant logic value (0 or 1), that is, the edge is redundant, then the circuit contains redundancy; otherwise, the circuit is nonredundant. Redundancy can be classified into two types: *observable* and *nonobservable*. Let us assume that an edge in the directed graph corresponding to a circuit is redundant. If the response of the original circuit for an input pattern differs from that of the reduced circuit obtained by eliminating the literal corresponding to the edge, the redundancy associated with the edge is known to be observable. On the other hand, if the original and the reduced circuit produce the same response for all input patterns, the redundancy associated with the re-placed edge is nonobservable.

Redundancy can also categorized as *serial* or *parallel*. If an edge feeds an AND or a NAND gate, and it can be replaced by a logic 1, the redundancy associated with the edge is known to be serial. Alternatively, if an edge feeds a

OR, a NOR, an EX–OR, or an EX–NOR gate and can be replaced by a logic 0, the associated redundancy is said to be parallel. It should be mentioned, however, that an edge can be redundant without necessarily being either serial or parallel.

Figures 3.16(a) and (b) show a gate-level circuit and the corresponding directed graph, respectively. The edge from vertex 4 to vertex 5 can be replaced by a 0, which is the same as the removal of the edge. The resulting graph is shown in Fig. 3.16(c). The removal of gate 4 edge in turn removes gate 4 from the circuit. Because the removal of gate 4 has no effect on the operation of the circuit, this gate is redundant. Also, the in-degree of vertex 5 is reduced from 3 to 2. The removal of a vertex or an edge from a directed graph is similar to the elimination of a redundant product term or a redundant literal, respectively, in the logic minimization process.

The random pattern testability of a circuit in some cases can actually be increased by incorporating observable redundancy into the circuit [3.9]. To illustrate, let us consider a two-input combinational circuit specified by the Karnaugh

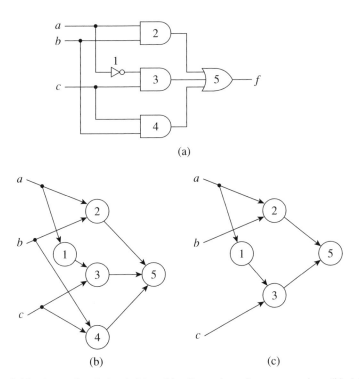

(a)

(b)   (c)

Figure 3.16   A gate-level circuit (a) and its directed graph representations (b), (c)

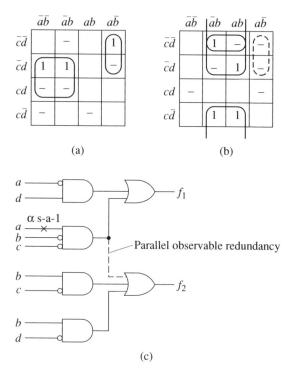

Figure 3.17   (a) f_1; (b) f_2; (c) a testable circuit with built-in redundancy

maps in Figs. 3.17(a) and (b). The implementation of the circuit is shown in Fig. 3.17(c). The inclusion of the parallel observable redundancy increases the probability of detection of *random pattern resistant* faults (i.e., faults that can only be detected by a single randomly generated test pattern). For example, in Fig. 3.17(c) if the observable redundancy is not included, the fault α stuck-at-1 can only be detected by the test pattern $abcd = 0000$. The probability of generating this particular pattern by a pseudo-random number generator is 1/16. If the redundant term $\overline{ab}c$ is included in function f_2, the fault α stuck-at-1 can be detected by $abcd = 000$-, that is, input d is don't care. Thus, the probability of generating a pattern to detect this fault is increased to 1/8. In other words, the inclusion of redundancy has enhanced the probability of detecting the fault, and it reduces the test generation time. The inclusion of the redundant signal line will obviously require test generation for stuck-at-0/1 faults on the line, thereby increasing the number of stuck-at faults to be considered.

3.6   Path Delay Fault Testable Combinational Logic Design

A two-level OR–AND circuit can be made fully testable for all path delay faults provided the following conditions are satisfied [3.10]:

1. The effect of a path delay fault can be propagated to an output via a single path.
2. There are no dynamic hazards on the propagating path when a two-pattern robust test is applied.

Such a test is called a *single path propagating hazard-free robust test* (SPP–HFRT). A two-pattern robust test detects a target fault independent of the presence of other delay faults [3.11]. For example, the two pattern test $\langle 010, 011 \rangle$ in Fig. 3.18 is a SPP–HFRT for a rising transition at the output of the path c–e–g.

A procedure for modifying a two-level OR–AND circuit such that all path delay faults in the circuit are testable by SPP–HFRTs has been proposed in Ref. 3.10. The original product-of-sums expression must be irredundant. Also, for each sum term in the expression, there exist *0-vertices* that are adjacent, in each and every literal in the sum term, to *1-vertices* of the function; a vertex (a combination of input variables) is called a 0 (1)-vertex if the function assumes the value of 0 (1) for the vertex. The transformation process consists of the following steps:

1. Identify the input variable x and the corresponding sum term p for the single untestable path in the circuit.
2. Identify a sum term q that contains the input variable (the existence of such a sum term is guaranteed). Next, use the dual transformation of the basic algebraic transformation $ab + ac = a(b + c)$, that is, $(a + b)(a + c) = a + bc$, to combine p and q.

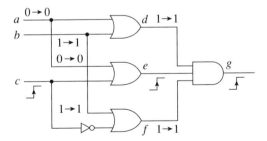

Figure 3.18   An OR–AND circuit

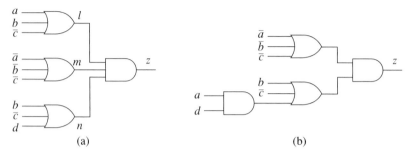

Figure 3.19   Path delay fault testable circuit implementation

To illustrate, let us consider the circuit shown in Fig. 3.19(a) in which the path b–n–z is not testable for rising transition at input b. The sum term associated with the path is $(b + \bar{c} + d)$. Another sum term that contains the input variable b is $(a + b + \bar{c})$. The two sum terms can be combined as follows:

$$(a + b + \bar{c})(b + \bar{c} + d) = b + (\bar{c} + d)(a + \bar{c}) = b + \bar{c} + ad.$$

Thus, the circuit of Fig. 3.19(a) can be transformed into that of Fig. 3.19(b). There exists a SPP–HFRT for each path in this circuit.

3.7   Testable PLA Design

The PLA logic structure is shown in Fig. 3.20(a). The input lines to the AND array are called *bit lines*, and the outputs of the AND array are called *word lines* or *product lines*. The input signals and their complements enter the AND array and are selectively connected to product lines in such a way that certain combinations of input variables produce a logic 1 signal on one or more product lines. The product lines are input to the OR array; hence, the outputs of the OR array are the sum-of-products form of Boolean functions of the PLA inputs. A cross-point b_k, p_k is marked (●) if the product term p_k depends on b_k, and a cross-point z_j, p_k is marked if p_k is a term of z_j. Figure 3.20(b) shows an example of a PLA realizing the functions

$$z_1 = x_1 x_2 + \bar{x}_1 \bar{x}_2,$$

$$z_2 = \bar{x}_1 x_2.$$

Three kinds of fault can normally occur in PLAs: stuck-at-faults, bridging faults, and cross-point faults [3.12–3.14]. A *cross-point* or *contact fault* may arise

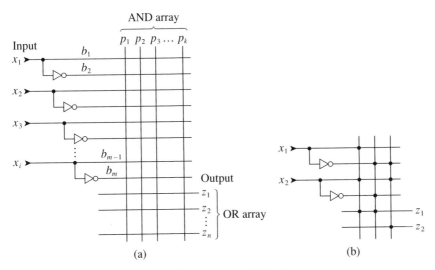

Figure 3.20 Programmable logic array

due to a missing contact on the cross-point of the AND and the OR array where there should be a contact, or due to an additional contact at the cross-point where there should not be a contact. A missing contact at a cross-point in the AND array, called a *growth* fault, causes an input variable to disappear from a product term. The resulting product term grows, that is, covers more minterms. On the other hand, a spurious contact at a cross-point in the AND array, called a *shrinkage* fault, can shrink a product term, that is, it covers fewer minterms. An extra contact in the OR array, an *appearance* fault, causes the erroneous inclusion of a product term in an output function. Similarly, a missing contact in the OR array, a *disappearance* fault, can eliminate a product term from an output function.

Several test generation algorithms have been proposed for PLAs [3.12–3.17]. The resulting test patterns can detect a large percentage of all single stuck-at, single bridging, and single cross-point faults. However, the testing problem for large PLAs is extremely complicated; a typical large PLA may have as many as 50 inputs, 67 outputs, and 180 product terms [3.18]. Thus, PLAs are augmented so that they become easily testable.

Hong and Ostapko [3.19] and Fujiwara [3.20] have independently proposed testable design techniques for PLAs. The resulting PLAs can be tested using a test set that is independent of the personality matrix of the PLA. Figure 3.21 shows the easily testable design for the NOR–NOR PLA implementing the following three-variable Boolean functions:

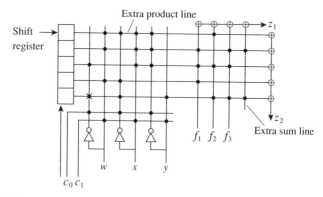

Figure 3.21 Fujiwara's approach for easily testable PLA design

$$f_1 = \overline{w}x + w\overline{x}y + \overline{w}\,\overline{x}y,$$

$$f_2 = wx\overline{y} + w\overline{x}y,$$

$$f_3 = \overline{w}\,\overline{x}y + w\overline{x}y + wx\overline{y}.$$

As can be seen in Fig. 3.21, a shift register, an extra product line, a parity checker circuit, and an additional output line have been incorporated to make the PLA easily testable. The shift register is loaded with a pattern consisting of a single 1 and the rest 0s. The shift register bit position with a 1 in it enables the product term connected to it; all other product terms remain deactivated. The extra product line is added to make the number of devices in bit lines odd. Similarly, the extra output line is used to make the number of devices in output lines odd. The parity check circuit checks the parity of the test responses. The extra control input lines in the decoders allow the selection of any bit line. This feature enables unique

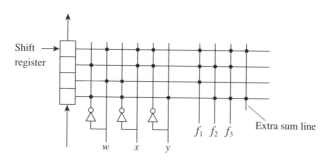

Figure 3.22 Easily testable PLA design using Khakbaz's approach

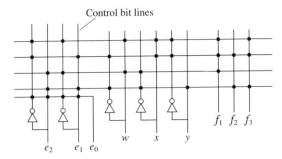

Control bit lines

f_1 f_2 f_3

e_2 e_1 e_0 w x y

Figure 3.23   Easily testable PLA design by Bozorgui-Nesbat *et al.*

identification of every cross-point. Thus, the test patterns and their responses are independent of the function of the PLA.

Khakbaz [3.21] modified Fujiwara's design in such a way that the parity checkers are eliminated. The easily testable version of the PLA specified by the previous equations is shown in Fig. 3.22. Although this design has less hardware overhead, the test patterns depend on individual PLAs.

Bozorgui-Nesbat *et al.* [3.22] replaced the shift register in Khakbaz's design by adding control bit lines as shown in Fig. 3.23. In general, the overhead due to these control lines is less than that resulting from using a shift register; however, the test pattern generation is more complicated than that in Khakbaz's approach.

References

3.1   Reddy, S. M., "A design procedure for fault-locatable switching circuits," *IEEE Trans. Comput.*, 1421–1426 (December 1972).

3.2   Dandapani, R., and S. M. Reddy, "On the design of logic networks with redundancy and testability considerations," *IEEE Trans. Comput.*, 1139–1149 (November 1974).

3.3   Hachtel, G., R. Jacoby, K. Keutzer, and C. Morrison, "On the relationship between area optimization and multifault testability of multilevel logic," *Proc. Intl. Workshop on Logic Synthesis, Research Triangle Park, NC*, 1–21 (1989).

3.4   Brayton, R., and C. McMullen, "The decomposition and factorization of Boolean expressions," *IEEE Intl. Symp. on Circuits and Systems*, 49–54 (May 1982).

3.5   Hatchel, G., "On properties of algebraic transformations," *IEEE Trans. on CAD*, 313–321 (March 1992).

3.6 Rajski, J. R., and J. Vasudevamurthy, "The testability preserving concurrent decomposition and factorization of Boolean expressions," *IEEE Trans. on CAD*, 778–793 (June 1992).

3.7 Batek, M. J., and J. P. Hayes, "Test set preserving logic transformation," *Proc. Intl. Test Conf.*, 454–458 (1993).

3.8 Krasniewski, A., "Can redundancy enhance testability?" *Proc. Intl. Test Conf.*, 483–491 (1991).

3.9 Krasniewski, A., "Logic synthesis for efficient pseudoexhaustive testability," *Proc. Intl. Test Conf.*, 66–72 (1991).

3.10 Kundu, S., and A. K. Pramanick, "Testability preserving Boolean transformation for logic synthesis," *Proc. IEEE VLSI Test Symp.*, 131–138 (1993).

3.11 Lin, C. J., and S. M. Reddy, "On delay fault testing in logic circuits," *IEEE Trans. on CAD*, 694–703 (September 1987).

3.12 Ostapko, D., and S. J. Hong, "Fault analysis and test generation for programmable logic arrays," *IEEE Trans. Comput.*, 617–626 (September 1979).

3.13 Cha, C. W., "A testing strategy for PLAs," *Proc. 15th Design Automation Conf.*, 326–334 (1978).

3.14 Smith, J., "Detection of faults in programmable logic arrays," *IEEE Trans. Comput.*, 845–853 (November 1979).

3.15 Agarwal, V., "Multiple fault detection in programmable logic arrays," *Proc. Symp. Fault-Tolerant Computing*, 227–234 (1979).

3.16 Fujiwara, H., K. Kinoshita, and H. Ozaki, "Universal test set for programmable logic arrays," *Proc. Symp. Fault-Tolerant Computing*, 137–142 (1980).

3.17 Bose, D., and J. Abraham, "Test generation for programmable logic arrays," *Proc. 19th Design Automation Conf.*, 574–580 (1982).

3.18 Law, H. F., and M. Shoji, "PLA design for the BELLMAC-32A microprocessor," *Proc. Intl. Conf. on Circuits and Components*, 161–164 (1982).

3.19 Hong, S. J., and D. L. Ostapko, "FIT-PLA: A programmable logic array for function independent testing," *Proc. Symp. Fault-Tolerant Computing*, 131–136 (1980).

3.20 Fujiwara, H., "A new PLA design for universal testability," *IEEE Trans. Comput.*, 745–750 (August 1984).

3.21 Khakbaz, J., "A testable PLA design with low overhead and high fault coverage," *IEEE Trans. Comput.*, 743–745 (August 1984).

3.22 Bozorgui-Nesbat, S., and E. J. McCluskey, "Lower overhead design for testability of programmable logic arrays," *IEEE Trans. Comput.*, 379–383 (April 1986).

Chapter 4 | Test Generation for Sequential Circuits

A logic circuit the output of which depends not only on the present values of the inputs but also on the past value of the inputs is known as a *sequential circuit*. The mathematical model of a synchronous sequential circuit is usually referred to as a *sequential machine* or a *finite state machine*. Henceforth, a synchronous sequential circuit will be referred to as a sequential circuit. Figure 4.1 shows the general model of a synchronous sequential circuit. As can be seen from the diagram, sequential circuits are basically combinational circuits with *memory* to remember past inputs. The combinational part of the circuit receives two sets of input signals: *primary* (coming from the external environment) and *secondary* (coming from the memory elements). The particular combination of secondary input variables at a given time is called the *present state* of the circuit; the secondary input variables are also known as *state variables*. If there are m secondary input variables in a sequential circuit, then the circuit can be in any one of 2^m different present states. The outputs of the combinational part of the circuit are divided into two sets. The primary outputs are available to control operations in the circuit environment, whereas the secondary outputs are used to specify the *next state* to be assumed by the memory.

4.1 Testing of Sequential Circuits as Iterative Combinational Circuits

Test generation for sequential circuits is extremely difficult because the behavior of a sequential circuit depends both on the present and on the past inputs. It takes an entire sequence of inputs to detect many of the possible faults in a sequential

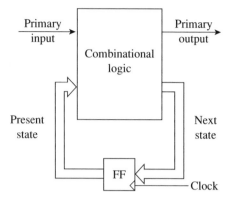

Figure 4.1 Synchronous sequential circuit

circuit. One of the earliest approaches for deriving tests for sequential circuits involves conversion of a sequential circuit into a one-dimensional array of identical combinational circuits. Most techniques for generating tests for combinational circuits can then be applied.

Figure 4.2 shows how n copies of the same sequential circuit are interconnected so that the state of the first is communicated to the second, the state of the second copy is communicated to the third, and so forth. In the diagram $x(i)$, $y(i)$, and $z(i)$, associated with the ith copy of the circuit, correspond to the input, the next state, and the output respectively, of the sequential circuit at the ith instant of time.

The circuit of Fig. 4.2 is a combinational network, because it consists of a cascade of identical *cells,* each having a pair of inputs (to receive input signals

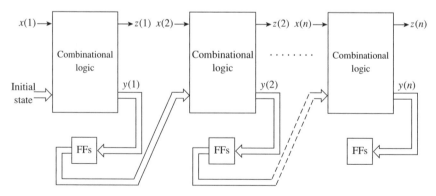

Figure 4.2 Iterative model of the sequential network

and information from its neighbor cell on the left) and a pair of outputs (to supply information to its neighbor on the right and to the network outputs). Thus, the transformation indicated in Fig. 4.2 replaces the problem of testing synchronous sequential circuits by the problem of finding tests for the iterative combinational network.

Each fault in the sequential network leads to n identical faults in the iterative network, and so the problem of fault detection in a synchronous sequential circuit can be treated as the problem of multiple-fault detection in a combinational network. The increased number of faults to be considered and the extra complexity of the replicated logic make this technique unrealistic for complex circuits. Moreover, it is necessary to derive test sequences that can detect a fault independently of the initial state of a circuit. This requires derivation of test sequences for all possible initial states, and then constructing a sequence that will detect the fault in all cases. However, this may not be computationally feasible.

4.2 State Table Verification

In this approach, a sequential circuit is tested by performing an *experiment* on it, that is, by applying an input signal and observing the output [4.1,4.2]. Hence, the testing problem may be stated as follows: Given the state table of a sequential machine, find an input/output sequence pair (X, Z) such that the response of the machine to X will be Z if and only if the machine is operating correctly. The application of this input sequence X and the observation of the response, to see if it is Z, is called a *checking experiment*; the sequence pair (X, Z) is referred to as a *checking sequence*.

Checking experiments are classified either as "adaptive" or "preset." In adaptive experiments, the choice of the input symbols is based on the output symbols produced by a machine earlier in the experiment. In preset experiments, the entire input sequence is completely specified in advance. A measure of efficiency of an experiment is its *length,* which is the total number of input symbols applied to the machine during the execution of an experiment. The derivation of *checking sequence* is based on the following assumptions:

1. The network is fully specified and deterministic. In a deterministic machine, the next state is determined uniquely by the present state and the present input.
2. The network is strongly connected, that is, for every pair of states q_i and q_j of the network, there exists an input sequence that takes the network from q_i to q_j.

3. The network in the presence of faults has no more states than those listed in its specification. In other words, no fault will increase the number of states.

To design checking experiments it is necessary to know the *initial state* of the network, which is determined by a *distinguishing* or a *homing sequence*.

An input sequence is said to be a *homing sequence* for a machine if the machine's response to the sequence is always sufficient to determine uniquely its final state. For an example, consider the machine of Fig. 4.3. It has a homing sequence 101, for, as indicated in Fig. 4.4, each of the output sequences that might result from the application of 101 is associated with just one final state. A homing sequence need not always leave a machine in the same final state; it is only necessary that the final state can be identified from the output sequence.

A *distinguishing sequence* is an input sequence that, when applied to a machine, will produce a different output sequence for each choice of initial state. For example, 101 is also a distinguishing sequence for Machine M of Fig. 4.3. As shown in Fig. 4.4, the output sequence that the machine produces in response to 101 uniquely specifies its initial state. Every distinguishing sequence is also a homing sequence because the knowledge of the initial state and the input sequence is always sufficient to determine uniquely the final state as well. On the other hand, not every homing sequence is a distinguishing sequence. For example, the machine of Fig. 4.5(a) has a homing sequence 010. As shown in Fig. 4.5(b), the output sequence produced in response to 010 uniquely specifies the final state of Machine N but cannot distinguish between the initial states C and D. Every reduced sequential machine possesses a homing sequence, whereas only a limited number of machines have distinguishing sequences.

At the start of an experiment a machine can be in any of its n states. In such a case, the *initial uncertainty* regarding the state of the machine is the set that

Present state	Input $x = 0$	$x = 1$
A	C,1	D,0
B	D,0	B,1
C	B,0	C,1
D	C,0	A,0

Next state, output

Figure 4.3   State table of Machine M

Initial state	Output sequence	Final state
A	0 0 1	C
B	1 0 0	A
C	1 0 1	B
D	0 1 1	C

Figure 4.4   Response of Machine M to homing sequence 101

contains all the states of the machine. A collection of states of the machine that is known to contain the present state is referred to as the *uncertainty*; the *uncertainty* of a machine is thus any subset of the state of the machine. For example, Machine M of Fig. 4.3 can initially be in any of its four states; hence, the initial uncertainty is $(ABCD)$. If an input 1 is applied to the machine, the successor uncertainty will be (AD) or (BC) depending on whether the output is 0 or 1, respectively. The uncertainties $(C)(DBC)$ are the 0-successors of $(ABCD)$. A *successor tree*, which is defined for a specified machine and a given initial uncertainty, is a structure that displays graphically the x_i-successor uncertainties for every possible input sequence x_i.

A collection of uncertainties is referred to as an *uncertainty vector*; the individual uncertainties contained in the vector are called the *components* of the vector. An uncertainty vector the components of which contain a single state each is said to be a *trivial uncertainty vector*. An uncertainty vector the components of which contain either single states or identical repeated states is said to be a *homogeneous uncertainty vector*. For example, the vectors $(AA)(B)(C)$ and $(A)(B)(A)(C)$ are homogeneous and trivial, respectively.

A homing sequence is obtained from the homing tree; a homing tree is a

Present state	Input $x = 0$ $x = 1$	Present state	Response to 010	Final state
A	B,0 D,0	A	0 0 0	A
B	A,0 B,0	B	0 0 1	D
C	D,1 A,0	C	1 0 1	D
D	D,1 C,0	D	1 0 1	D
	(a)		(b)	

Figure 4.5   (a) Machine N; (b) response of Machine N to 101

successor tree in which a node becomes terminal if one of the following conditions occurs:

1. The node is associated with an uncertainty vector the nonhomogeneous components of which are associated with the same node at a preceding level.
2. The node is associated with a trivial or a homogeneous vector.

The path from the initial uncertainty to a node in which the vector is trivial or homogeneous defines a homing sequence.

A distinguishing tree is a successor tree in which a node becomes terminal if one of the following conditions occurs:

1. The node is associated with an uncertainty vector the nonhomogeneous components of which are associated with the same node at a preceding level.
2. The node is associated with an uncertainty vector containing a homogeneous nontrivial component.
3. The node is associated with a trivial uncertainty vector.

The path from the initial uncertainty to a node associated with a trivial uncertainty defines a distinguishing sequence. As an example, the homing sequence 010 is obtained as shown in Fig. 4.6 by applying the terminal rules to Machine N of Fig. 4.5. The derivation of the distinguishing sequence 101 for Machine M of Fig. 4.3 is shown in Fig. 4.7.

During the design of checking experiments it is often necessary to take the machine into a predetermined state, after the homing sequence has been applied. This is done with the help of a *transfer sequence*, which is the shortest input sequence that takes a machine from state S_i to state S_j. The procedure is an

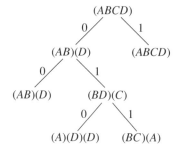

Figure 4.6   Homing tree for Machine N

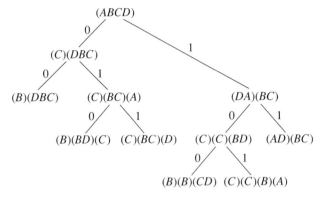

Figure 4.7 Distinguishing tree for Machine M

adaptive one, because the transfer sequence is determined by the response of the homing sequence. As an example, let us derive the transfer sequence that will take Machine M of Fig. 4.3 from state *B* to state *C*. To accomplish this, we assume that the machine is in state *B*. We form the transfer tree as shown in Fig. 4.8; it can be seen from the successor tree that the shortest transfer sequence that will take the machine from state *B* to state *C* is 00.

Instead of a homing sequence, a *synchronizing sequence* may be used at the beginning of the checking experiment; it takes a machine to a specified final state regardless of the output or the initial state of the machine. Some machines possess synchronizing sequences—others do not. For a machine, one can construct a synchronizing tree by associating with each node the uncertainty regarding the final states (regardless of the output) that results from the application of the input symbols. For example, if the initial uncertainty of Machine P of Fig. 4.9(a) is (*ABCD*), the 0-successor uncertainty is (*ABC*); it is not necessary to write down

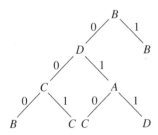

Figure 4.8 Transfer tree

Present state	Input	
	$x = 0$	$x = 1$
A	B,1	C,0
B	A,0	D,1
C	B,0	A,0
D	C,1	A,1

Next state, output

(a)

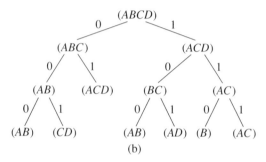

(b)

Figure 4.9   (a) Machine P; (b) synchronizing tree for Machine P

the repeated entries (*BABC*). A node becomes terminal whenever either of the following conditions occurs:

1. The node is associated with an uncertainty that is also associated with the same node at a preceding level.
2. A node in the nth level is associated with an uncertainty containing just a single element.

The synchronizing tree for Machine P of Fig. 4.9(a) is shown in Fig. 4.9(b). A synchronizing sequence is described by a path in the tree leading from the initial uncertainty to a singleton uncertainty. For Machine P, 110 is the synchronizing sequence that, when applied to the machine, synchronizes it to state *B*.

Designing Checking Experiments

Basically, the purpose of a checking experiment is to verify that the state table of a finite state machine accurately describes the behavior of the machine. If during the execution of the experiment the machine produces a response that is different from the correctly operating machine, the machine is definitely faulty.

Such experiments can be used only to determine whether or not *something* is wrong with a machine; it is not possible to conclude from these experiments *what* is wrong with the machine.

We will now show how to design a checking experiment for any strongly connected, *diagnosable* sequential machine. A *diagnosable machine* has the property of having at least one distinguishing sequence. The checking experiment can be divided into three phases:

1. *Initialization phase.* During the initialization phase, the machine under test is taken from an unknown initial state to a fixed state. A reduced, strongly connected machine can be maneuvered into some fixed state *s* by the following method:
   (a) Apply a homing sequence to the machine and identify the current state of the machine.
   (b) If the current state is not *s*, apply a transfer sequence to move the machine from the current state to *s*.

   It is apparent that the initialization phase is adaptive; however, it can be preset if a machine has a synchronizing sequence.

2. *State identification phase.* During this phase, an input sequence is applied so as to cause the machine to visit each of its states and display its response to the distinguishing sequence.

3. *Transition verification phase.* During this phase, the machine is made to go through every state transition; each state transition is checked by using the distinguishing sequence.

Although these three phases are distinct, in practice the subsequences for state identification and transition verification are combined whenever possible, in order to shorten the length of the experiment.

The state table verification approach can be considered as a form of functional testing. The major limitation of this technique is that it results in excessively long test sequences, and it is therefore only of theoretical interest. However, the concept of employing state tables for test generation and testability enhancement of sequential circuits has been used in several recent techniques (see Sec. 4.5).

4.3   Test Generation Based on Circuit Structure

Several test generation techniques for sequential circuits have been proposed in recent years. In this section we will consider some of these techniques that generate test sequences from the circuit structure. Agarwal *et al.* [4.3] have proposed a simulation-based approach called *CONTEST* for sequential circuit test generation. It uses a concurrent fault simulator that allows it to generate tests for a group of faults. An initialization vector to bring the fault-free circuit to a known state is derived first using the fault simulator. For each input vector, a cost function defined as the number of flip-flops that are in an unknown state is computed. This vector is modified till the cost becomes zero. After the circuit has been initialized, an input vector is applied and its cost, defined as the number of logic gates on the path from the effect of a fault to any primary output, is calculated. The input vector is modified by the fault simulator till the effect of a fault reaches a primary output, that is, until the cost of the input vector is zero. The input vector is then included in the set of test vectors. For test generation of a single fault, the weighted sum of costs for two objectives—sensitization of the fault and the propagation of the fault—is considered as the cost function. The cost of the first objective is the dynamic controllability measure computed from the current circuit state determined by the simulator. The cost function for the second objective is the dynamic observability measure of the fault through all paths from the fault site to the primary outputs. The minimizations of the first cost and the second cost lead to the activation of the fault and subsequent selection of test vectors for detecting the fault, respectively.

Ghose *et al.* [4.4] have proposed a test generation technique based on the concept of path sensitization used in combinational circuit test generation algorithms, for example, PODEM. This technique takes into account both the structure and the state table of a sequential circuit. It is assumed that the circuit under test has a *reset state*. A test sequence is applied to the circuit with the reset state as the starting state. The test generation process consists of the following steps:

1. Generate a test vector for the assumed single stuck-at fault such that the effect of the fault is propagated to the primary outputs or to the secondary outputs, that is, the outputs of the flip-flops. Each primary output as well as each secondary output is considered an independent output of a combinational circuit. A test vector for a fault is identified as an *excitation vector*, and the present state part of an excitation vector is called the *excitation state*.

2. Derive an input sequence to take the circuit from the reset state to the excitation state; this input sequence is called the *justification sequence*. Obviously, a justification sequence is not necessary if the excitation state part of a test vector is the reset state. If the effect of a fault can be propagated to the primary outputs by the derived test vector, and the justification sequence can take the faulty circuit from the reset to the excitation state, then the test vector is valid. However, if the test vector can propa-

gate the effect of the fault only to the outputs of the flip-flops, that is, the next state is different from the one expected, the following step is also necessary for successful test generation for the assumed fault.

3. Derive an input sequence such that the last bit in the sequence produces a different output for the fault-free and the faulty states of the circuit under test. Such an input sequence is called the *differentiating sequence*.

A test sequence for the fault under test is obtained by concatenating the justification sequence, the excitation vector, and the differentiating sequence. This test sequence is simulated in the presence of the fault to check if the fault is detected. If the fault is not detected, the differentiating sequence is not valid. Also, a valid differentiating sequence cannot be obtained if the fault-free and the faulty states are equivalent in the fault-free sequential circuit.

Let us illustrate the technique by deriving a test sequence for the fault α stuck-at-0 in the sequential circuit shown in Fig. 4.10(a); the state table of the fault-free circuit is shown in Fig. 4.10(b).

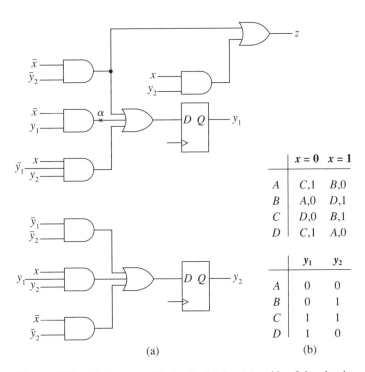

	$x = 0$	$x = 1$
A	$C,1$	$B,0$
B	$A,0$	$D,1$
C	$D,0$	$B,1$
D	$C,1$	$A,0$

	y_1	y_2
A	0	0
B	0	1
C	1	1
D	1	0

(a)   (b)

Figure 4.10   (a) A sequential circuit; (b) the state table of the circuit

A partial test vector for the fault is first derived:

$$x \quad y_1 \quad y_2$$
$$0 \quad 1 \quad -$$

The value of y_2 has to be chosen such that the effect of the fault is propagated to the primary output. If $y_2 = 0$, the fault is neither propagated to the output nor does it affect the next state. On the other hand, $y_2 = 1$ affects the next state variable Y_1. Therefore, the excitation vector for the fault is

$$x \quad y_1 \quad y_2$$
$$0 \quad 1 \quad 1$$

and the excitation state is 11 (i.e., state C).

Next, we derive the justification sequence that can take the fault-free machine from the reset state A to the excitation state C. It can be verified from the state table that the justification sequence needs to contain a single bit only, which is 0. Thus, the test sequence derived so far consists of

$$0 \quad 0$$

and the corresponding fault-free output sequence and the next states are

$$A \xrightarrow[1]{0} C \xrightarrow[0]{0} D$$

In the presence of the assumed stuck-at fault, the next state/output sequence is

$$A \xrightarrow[1]{0} C \xrightarrow[0]{0} A$$

Because the output sequence is the same as in the fault-free case, the fault is not detectable. However, the final state in the presence of the fault is A instead of expected D; in other words, the effect of the fault α propagated only to the outputs of the flip-flops. Therefore, a differentiating sequence that produces a different output sequence for state A and state D has to be concatenated with the previously derived test sequence. The differentiating sequence is derived as follows:

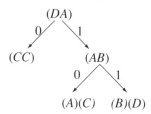

Either 10 or 11 can be used as the differentiating sequence. The choice of 11 results in the following test sequence for the assumed stuck-at fault in the circuit of Fig. 4.10(a):

$$0 \quad 0 \quad 1 \quad 1$$

The corresponding next state/output sequence for the fault-free circuit is

$$A \overset{0}{\underset{1}{\rightarrow}} C \overset{0}{\underset{0}{\rightarrow}} D \overset{1}{\underset{0}{\rightarrow}} A \overset{1}{\underset{0}{\rightarrow}} B$$

and for the faulty circuit

$$A \overset{0}{\underset{1}{\rightarrow}} C \overset{0}{\underset{0}{\rightarrow}} A \overset{1}{\underset{0}{\rightarrow}} B \overset{1}{\underset{1}{\rightarrow}} D$$

Thus, the fault is detected by the derived test sequence.

Kelsey *et al.* [4.5] have proposed a sequential circuit test generation algorithm called *FASTEST* that works strictly in forward time. In other words, all assignments are made at primary inputs; any assignment of test value that requires justification is avoided. It uses a nine-valued logic model, $0, 1, x, D, \overline{D}, G0$ (good zero), $G1$ (good one), $F0$ (faulty zero), and $F1$ (faulty one), proposed by Muth [4.6]. The nine-valued model takes into account the repeated effects of a fault during the sequential circuit test generation process.

The input to *FASTEST* consists of the description of the circuit under test, and a fault list for which the test sequence has to be derived. For each fault in the list, the *excitation time frame* and the *total time frame* are derived by using an algorithm called the *initial time frame algorithm*. The excitation time frame is the minimum number of time frames needed to excite a fault, and the minimum of time frames needed to observe the fault is the total time frame. Obviously, the total time frame cannot be smaller than the excitation time frame, because this might be the minimum number of time frames that are necessary to test the fault.

The initial time frame algorithm first initializes the excitation time frame and the total time frame to 1. If the selected fault is not excited in the current excitation time frame, the excitation time frame is incremented. This process is continued till the fault is excited, or the maximum number of time frames is utilized without success, in which case the fault is assumed to be untestable. Once the fault is excited, the total time frame is set equal to the excitation time frame. If the fault is not observable at some primary output, the total time frame is incremented to check whether the excited fault is observable. In case the fault is still not observable, the total time frame is incremented till either the fault is observable or the maximum number of time frames has been used. If the total time frame reaches the maximum time frame, a test is performed to determine whether the excitation time frame is equal to the maximum time frame. The fault is undetectable if the test is successful. Otherwise, the excitation time frame is incremented and the algorithm is continued.

Once the initial time frame algorithm provides *FASTEST* with the number of excitation time frames for a fault and the total number of time frames, the test generation algorithm determines an initial objective to excite the fault and propagates it to an output. The nine-valued logic model is used for this purpose. The initial objective is *G1* (*G0*) if the fault is stuck-at-0 (stuck-at-1). The initial objective for propagating the faulty value depends on the gate type and the faulty value. If it is not possible to move the circuit to a state closer to detecting the fault, the primary input value assigned last is complemented (assuming the complement value has not been tried).

The test generation algorithm starts backtracing after an initial objective has been determined. This is continued until a primary input or a flip-flop is reached in the first time frame. The fault is untestable if a flip-flop is reached, because it is not possible to assign value to a flip-flop in the first time frame. If a primary input is reached during the backtrace process, it is assigned a value that is most likely to satisfy the initial objective. Next, the circuit is *implicated*, that is, the effect of the given values of primary inputs on the internal lines of the circuit is determined. If there is an inconsistency, different values are assigned to a set of primary inputs, and the circuit is reimplicated and checked for inconsistencies. The primary outputs assume a D or $\overline{D}$ if the reimplicated circuit is consistent; otherwise, the test generation process is restarted with a new initial objective.

4.4 Functional Fault Models

Test generation of sequential circuits based on the traditional stuck-at fault model is extremely complicated and time consuming because the effect of a stuck-at fault may propagate to the circuit output after several time frames. By modeling of faults at the state table, that is, the functional level, we can generate tests for a sequential circuit before the circuit is actually implemented.

Cheng and Jou [4.7] have shown that tests generated at the state table level produce high fault coverage of stuck-at faults at the gate level. They have proposed a functional fault model, called the *single transition fault model*, to model the faults in sequential circuits. In this model, any stuck-at fault in a circuit is assumed to result in an erroneous state transition. In other words, a fault may

	$x = 0$	$x = 1$
A	C,0	B,1
B	C,1	A,0
C	A,0	B,0

Figure 4.11   A state table

force the sequential circuit to move from a present state to a next state other than that specified in the state table of the circuit. To illustrate, let us consider the state table shown in Fig. 4.11. An implementation of the state table assuming the state assignment $y_1 y_2 = 10$, 11, and 00 for state A, B, and C, respectively, is shown in Fig. 4.12. Let us next assume that α is stuck-at-0 in Fig. 4.12. The state table corresponding to the faulty circuit is shown in Fig. 4.13. As can be seen in the state table, the destination states for present states A and C when input $x = 1$ have changed (from expected B to A). Note that there is no change in the output value.

The main problem with the single transition fault model is that the number of faults to be considered in an n-state sequential circuit will be equal to $n \times (n - 1) \times 2^x$, where x is the number of inputs to the circuit. For example, the number of possible single transition faults in the state table of Fig. 4.13 is $12 (= 3 \times 2 \times 2)$.

Pomeranz and Reddy [4.8] have proposed a fault model that, unlike the single state transition model, considers only *state table faults* that result from the presence of stuck-at faults in the two-level implementation of the combinational logic part of a sequential circuit. Thus, composite states of the form S_1/S_2, where S_1 and S_2 belong to the fault-free and the faulty machine, respectively, are considered. For example, the state table of a sequential circuit with state table fault

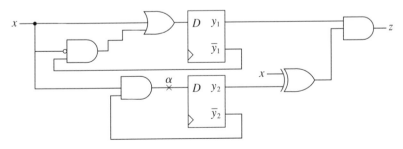

Figure 4.12   An implementation of the state table of Fig. 4.11

	$x = 0$	$x = 1$
A	C,0	A,1
B	C,1	A,0
C	A,0	A,0

Figure 4.13   Faulty state table

$A \xrightarrow{0} C/D$ is shown in Fig. 4.14. This fault forces the circuit to move from state A to state D instead of state C, when the input $x = 0$.

The number of faults under the state table fault model does not exceed the total number of stuck-at faults in the two-level circuits. However, a multilevel rather than a two-level combinational logic is more likely to be used in the actual implementation of a sequential circuit. Some of the stick-at faults in the multilevel logic may not be covered by the state table faults derived from the two-level implementation.

4.5   Test Generation Based on Functional Fault Models

Cheng and Jou [4.7] have proposed a test generation technique for sequential circuits that considers a selected subset of all single transition faults. The selected subset covers the remaining transition faults. The test procedure first initializes the circuit under test from a present state to an originating state for a faulty transition. After the circuit moves into a wrong destination state, a *state-group differentiating (SGD)* sequence is applied to differentiate between the expected state and the faulty state. A reduced n-state sequential circuit has a collection of $n - 1$ SGD sequences, each of which distinguishes between a state s and one other state in the circuit; the SGD sequences form a *SGD sequence set* for state

	$x = 0$	$x = 1$
A	C/D,0	B,0
B	D,0	C,0
C	A,1	D,1
D	B,0	A,1

Figure 4.14   State table with faulty and fault-free transitions

	$x = 0$	$x = 1$
A	B,0	C,0
B	C,0	D,1
C	A,1	D,1
D	B,0	A,1

Figure 4.15 State table

s. The state table shown in Fig. 4.15 does not have a distinguishing sequence. The input $x = 1$ can distinguish between states A and states B, C, and D, and the input $x = 0$ can distinguish between state A and state C. The SGD sequences $T(A, C) = 0$ and $T(A, \{B, C, D\}) = 1$ constitute a SGD sequence set for state A. Similarly, sequences $T(B, C) = 0$ and $T(B, A) = 1$ form the SGD sequence set for state B; sequences $T(C, \{A, B, D\}) = 0$ and $T(C, A) = 1$ form a SGD sequence set for state C; and sequences $T(D, C) = 0$ and $T(D, A) = 1$ form a SGD sequence set for state D.

Pomeranz and Reddy [4.8] have proposed an algorithm, based on state table fault model, that generates a minimal length test sequence for a fault in a sequential circuit. Both the fault-free and the faulty circuits are used during the test generation process; it is assumed that both circuits can be initialized to known *reset* states, which may be different. Tests are generated by simultaneously transferring the fault-free and the faulty circuits from the reset state to a *fault-detection state*. A composite state R_1/R_2 is a fault-detection state if different output sequences are obtained in response to an input sequence applied to the fault-free circuit in state R_1 and the faulty circuit in state R_2. In other words, the applied input sequence is a distinguishing sequence for states R_1 and R_2.

A composite state Q_1/Q_2 is considered to be *reachable* from another composite state P_1/P_2 if there exists an input sequence that can transfer the circuit from P_1/P_2 to Q_1/Q_2. The *reachability index* of a composite state is assigned a value of 1 if this is reachable at time i starting from the initial composite state S_0/S_0^f at time 0; otherwise, the reachability index is 0. The reachability indices of all states in a sequential circuit can be calculated using a procedure that we illustrate using the sequential circuit of Fig. 4.16(a). Let us assume that the initial state of each of the fault-free circuit and the faulty circuit is A. At time $i = 0$, only the initial state has a reachability index of 1; all other states have reachability indices of 0. The next state entries for A corresponding to $x = 0$ and $x = 1$ are C/D and B, respectively. Therefore, at time $i = 1$ the reachability index for C/D is 1, and that

	x = 0	x = 1
A	C/D,0	B,0
B	D,0	C,0
C	A,1	D/A,1
D	B,0	A,1

(a)

i	A	B	C	D	C/D	A/B	D/A
0	1	0	0	0	0	0	0
1	0	1	0	0	1	0	0
2	0	0	1	1	0	1	1

(b)

Figure 4.16 (a) Sequential circuit; (b) reachability indices for the sequential circuit in (a)

for B is also 1. In other words, only states B and A/B are reachable at time $i = 1$; all other states are unreachable. At $i = 2$, states D and C can be reached from state B, and states A/B and D/A can be reached from C/D. Thus, both A/B and D/A have reachability indices of 1 in Fig. 4.16(b). Note that D/A is a fault detection state. Therefore, it is not necessary to derive additional reachability indices. Any state, not necessarily the reset state, can be used for computing reachability indices. Also, if no fault detection state is found while computing reachability indices, tests cannot be derived for the state table faults.

Once the reachability indices of a sequential circuit have been derived, the test sequence for the sequential circuit is generated by setting the state of the circuit to the fault detection state R_1/R_2 that was reached at time t. An input is applied such that the output response obtained when the circuit is at R_1 is different from that obtained when the circuit is at R_2. Next, a predecessor state of R_1/R_2, for example, Q_1/Q_2, is identified at time unit $t - 1$. The input value required to move the circuit from Q_1/Q_2 to R_1/R_2 is obtained from the state table. Predecessor states and input values are derived in a similar manner till the time unit is 0. At time unit 0, the reset state is reached, and the test sequence generation process is completed.

We illustrate the test sequence generation process for the state table fault $C \xrightarrow{1} D/A$ in Fig. 4.16(a). It can be verified from Fig. 4.16(a) that D/A is a fault-detection state; other fault-detection states are $A/C, A/D, B/C, B/D, C/A, C/B, C/D,$ $D/B,$ and D/C. Figure 4.16(b) indicates that at time unit $i = 2$ the index of the fault detection state D/A is 1. An input of 1 at state D/A produces different outputs for the faulty and the fault-free circuits. The only predecessor state of D/A having an index 1 at time $i = 1$ is C/D. The required input to move from C/D to D/A is 1. The lone predecessor state of C/D having an index 1 at time $i = 1$ is C/D. The required input to move from C/D to D/A is 1. The sole predecessor state of C/D

having index 1 at time $i = 0$ is A; the input necessary to move from A to C/D is 0. Therefore, the minimum length of a test sequence is 3, and the sequence is

$$A \xrightarrow[0]{0} C/D \xrightarrow[1]{1} D/A \xrightarrow[1/0]{1} A/B$$

Next, let us derive the test sequence for the state table fault $A \xrightarrow{0} C/D$. From Fig. 4.16(b) it can be seen that C/D is a fault detection state with an index 1 at time $i = 1$. An input of 0 distinguishes between the fault-free state C and the faulty state D. The only predecessor state of C/D at time $i = 0$ is A. Input 0 changes state A to C/D. Thus, the length of the test sequence needed to detect the state table fault $A \xrightarrow{0} C/D$ is 2; the sequence is

$$A \xrightarrow{0} C/D \xrightarrow[1/0]{0} A/B$$

As mentioned previously, this method for test sequence generation is based on the assumption that a single stuck-at fault in a sequential circuit will manifest itself as a state table fault. It has been suggested in Ref. 4.8 that equivalent state table faults for all single stuck-at faults in a sequential circuit can be extracted from the *pseudo-implementation* that is, two-level AND–OR implementation, of the combinational logic part of the sequential circuit. Because in practice multi-level logic is more likely to be used, realistic fault coverage can only be achieved by deriving state table equivalent faults of the stuck-at faults in the multilevel circuit. This state table fault derivation approach, although very simple in principle, may require unacceptably high simulation time.

Sheu and Lee [4.9] have proposed a technique for synthesizing sequential circuits that have built-in parity checking to detect single state transition faults. A state transition fault can only be detected by propagating its effect to the primary outputs. However, the effect of a fault cannot be detected at the outputs if the circuit produces the same output for the correct and the erroneous states and then moves to the identical next state. By encoding the states of a sequential circuit with a particular parity, it is possible to differentiate a correct state transition from an erroneous one. The state encoding process consists of the following steps:

1. Construct a *distinguishing table* from the state table of the sequential circuit.
2. Compute the *undistinguishability value* for each state pair in the distinguishing table.

3. Arrange the state pairs in the form of a list with decreasing order of un-distinguishable value; that is, a state pair with the largest undistinguisha-bility value is placed at the head of the list, followed by the pair with the next highest undistinguishable value, and so on.

To illustrate the state encoding process, let us consider the state table shown in Fig. 4.17. The associated distinguishable table is shown in Fig. 4.18(a). A state pair is indistinguishable if they produce the same outputs for the same inputs, although they may move to different next states. Each cell of the table corresponds to a state pair defined by the intersection of the row and the column headings. The distinguishability of a state pair is recorded by placing a $\times$ in the correspond-ing cell. If a state pair (p,q) is indistinguishable for an input, the resulting next state pair is entered in the cell (p,q).

For the state table of Fig. 4.17, the state pair (A,C) is *totally distinguishable* because states A and C move to different states with different outputs for both inputs. Also, state pairs (B,D) and (D,E) are totally distinguishable. States B and E move to the same next state B with output 0 when the input is 1, and they move respectively to states C and D with output 0 when the input is 0. Thus, the state pair (B,E) is *partially equivalent*, and it may become equivalent if (C,D) is equiv-alent. The entries in cell (B,E) are $*$ and (C,D).

The undistinguishability of a state pair (p,q) can be computed from the distin-guishable table, and it is

= number of next state pairs in cell (p,q)

+ number of (p,q) in the distinguishable table + xN,

where N is the number of states in the sequential circuit with L outgoing edges, and x $(1 \leq x \leq L)$ is the number of transitions from (p,q) that result in the same next state and the same output for an input. For example, undistinguishability of $(B,C) = 1 + 2 + 0 = 3$, because there is one state pair in cell (B,C), and (B,C) appears in cell (A,B) and cell (A,E); there is no transition from (B,C) with the same next state. The undistinguishability of $(B,E) = 1 + 1 + 5$, because there

	$x = 0$	$x = 1$
A	$B,1$	$C,0$
B	$D,0$	$B,0$
C	$A,0$	$D,1$
D	$E,1$	$B,1$
E	$C,0$	$B,0$

Figure 4.17   A state table

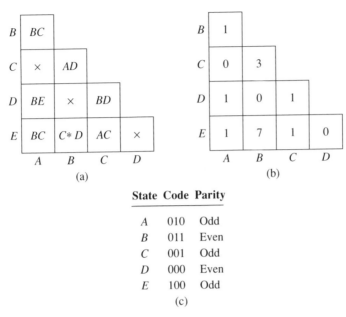

<table>
<tr><td>B</td><td>BC</td></tr>
</table>

B	BC			
C	×	AD		
D	BE	×	BD	
E	BC	C∗D	AC	×
	A	B	C	D

(a)

B	1			
C	0	3		
D	1	0	1	
E	1	7	1	0
	A	B	C	D

(b)

State Code Parity

State	Code	Parity
A	010	Odd
B	011	Even
C	001	Odd
D	000	Even
E	100	Odd

(c)

Figure 4.18   (a) Distinguishable table; (b) undistinguishable values; and (c) parity encoding of states

is one state pair in cell (B,E). Also, (B,E) appears only in cell (A,D), $x = 1$, and $N = 5$. The undistinguishability values for all states are shown in the table in Fig. 4.18(b). From this table, an ordered list of state pairs is obtained:

$$(B,E)(B,C)(A,B)(A,D)(A,E)(C,D)(C,E).$$

The state pairs (B,E) and (B,C) occupy the first and the second position in the list because they have undistinguishability value of 7 and 3, respectively; the rest of the state pairs have undistinguishability value of 1 and can be arranged in any order. The state pair (B,E) should get assigned to different parity first, followed by pair (B,C), and so on. A possible state encoding for the circuit, with states A, C, and E being assigned odd parity codes and states B and D being assigned even parity codes, is shown in Fig. 4.18(c).

A sequential circuit in which the parity codes are assigned as just discussed can be tested for single state transition faults by incorporating a parity checker as shown in Fig. 4.19. An input sequence is derived that makes a sequential circuit to go through all the states, starting from an initial state. The parity bits resulting from the application of this input sequence constitute the reference output sequence. A state transition fault will result in erroneous parity bits in the sequence, and it will therefore be detected.

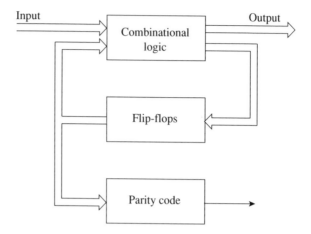

Figure 4.19   Parity-encoded sequential circuit

References

4.1 Kohavi, Z. *Switching and Finite Automata Theory*, Chap. 13, McGraw-Hill (1970).

4.2 Hennie, F. C., *Finite State Models for Logical Machines*, Chap. 3, John Wiley (1968).

4.3 Agrawal, V. D., K. T. Cheng, and P. Agrawal, "*CONTEST:* A concurrent test generator for sequential circuits," *Proc. 25th Design Automation Conf.*, 84–89 (1988).

4.4 Ghose, A., S. Devadas, and A. R. Newton, "Test generation and verification for highly sequential circuits," *IEEE Trans. on CAD*, 652–667 (May 1991).

4.5 Kelsey, T., K. K. Saluja, and S. Y. Lee, "An efficient algorithm for sequential test generation," *IEEE Trans. Comput.*, 1361–1371 (November 1993).

4.6 Muth, P., "A nine-valued circuit model for test generation," *IEEE Trans. Comput.*, 630–636 (June 1976).

4.7 Cheng, K. T., and J. Y. Jou, "Functional test generation for finite state machines," *Proc. Intl. Test Conf.*, 162–168 (1990).

4.8 Pomeranz, I., and S. M. Reddy, "On achieving complete fault coverage for sequential machines," *IEEE Trans. Comput.*, 378–385 (March 1994).

4.9 Sheu, M.-L., and C. L. Lee, "Simplifying sequential circuit test generation," *IEEE Design and Test*, 28–38 (Fall 1994).

Chapter 5 | Design of Testable Sequential Circuits

Several techniques for synthesizing testable combinational circuits have been proposed in recent years (see Chap. 3); however, not much has been reported on synthesis techniques for testable sequential circuits. As discussed before, test generation for sequential circuits is significantly more complex than that for combinational circuits. Many design guidelines have been proposed to improve the testability of sequential circuits by adding extra test points; however, these involve nonsystematic (i.e., ad hoc) design modifications and depend heavily on a designer's ingenuity. A well-established approach to solve testability problem for sequential circuits is to constrain the design in a way that provides direct access to the memory elements in a circuit. This is known as the *scan design methodology*, and it is widely used in designing sequential circuits with enhanced testability. In this chapter, we will discuss some of the ad hoc design rules, several variations of the scan methodology, and some other techniques for improving sequential circuit testability.

5.1 Controllability and Observability

There are two key concepts in designing for testability: *controllability* and *observability*. Controllability refers to the ability to apply test patterns to the inputs of a subcircuit via the primary inputs of the circuit. For example, in Fig. 5.1(a) if the output of the equality checker circuit is always in the state of *equal*, it is not possible to test whether the equality checker is operating correctly or not. If a control gate is added to the circuit (Fig. 5.1(b)), the input of the equality checker

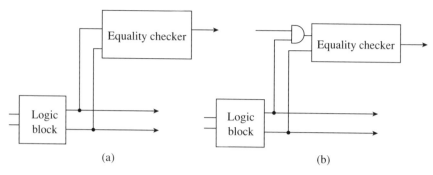

Figure 5.1   Controllability

and, hence, the operation of the circuit can be controlled. Therefore, to enhance the controllability of a circuit, the state that cannot be controlled from its primary inputs has to be reduced.

Observability refers to the ability to observe the response of a subcircuit via the primary outputs of the circuit or at some other output points. For example, in Fig. 5.2 the outputs of all three AND gates are connected to the inputs of the OR gate. A stuck-at-0 fault at the output of the AND gate 3 is not detectable because the effect of the fault is masked and cannot be observed at the primary output. To enhance the observability, we must observe the output of the gate separately as shown.

In general, the controllability/observability of a circuit can be enhanced by incorporating some control gates and input lines (controllability), and by adding some output lines (observability).

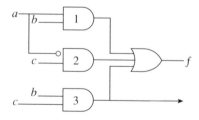

Figure 5.2   Observability

5.2   Ad Hoc Design Rules for Improving Testability

Ad hoc rules are used to improve testability of specific circuits. One of the simplest ways of achieving this is to incorporate additional control and observation points in a circuit. For example, the fault α stuck-at-1 in the circuit of Fig. 5.3(a) is undetectable at the circuit output. The addition of an extra output line in the circuit makes the fault detectable (Fig. 5.3(b)).

The usefulness of inserting a control point can be understood from the circuit shown in Fig. 5.4(a). The output of the NOR gate is always 0; therefore, it is not possible to determine whether the gate is functioning correctly or not. If a control point is added to the circuit as shown in Fig. 5.4(b), the NOR gate can be easily tested for single stuck-at fault.

Another way of improving testability is to insert multiplexers to increase the number of internal points that can be controlled or observed from the external pins. For example, in the circuit of Fig. 5.5(a) the fault α stuck-at-0 is undetectable at the primary output Z. By incorporating a multiplexer as shown in Fig. 5.5(b), input combination 010 can be applied to detect the fault via the multiplexer output.

A different way of achieving access to internal points is to use tristate drivers as shown in Fig. 5.6. A test mode signal could be used to put the driver into the high impedance state. In this mode, the internal point could be used as a control point. When the driver is activated, the internal point becomes a test point.

Another approach to improve testability is to permit access to a subset of the logic as shown in Fig. 5.7 [5.1,5.2]. Module B is physically embedded between the two modules A and C. A set of gates G and H is inserted into each of the inputs and outputs, respectively, of module B. In normal operation, the test control signal is such that modules A, B, and C are connected and the complete network performs its desired function. In the test mode, the test control input is changed;

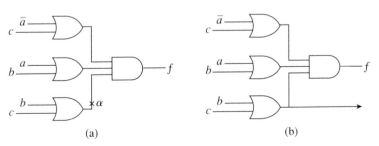

(a)   (b)

Figure 5.3   (a) Circuit with an undetectable fault; (b) addition of an extra output line

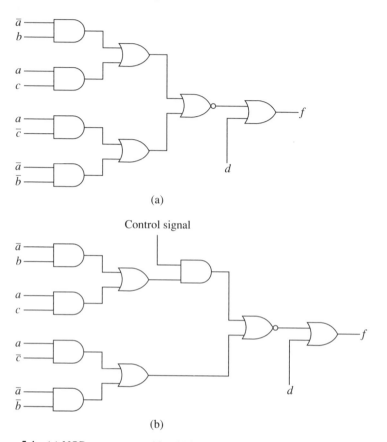

(a)

(b)

Figure 5.4   (a) NOR gate not testable; (b) improvement in testability using a control point

module B is connected to the primary inputs and outputs of the board. In this mode, the control signal also causes the outputs of module C to assume a high impedance state, and hence C does not interfere with the test results generated by B. Basically, this approach is similar to the previously discussed technique of using multiplexers to improve testability.

The test mode signals required by the added hardware such as multiplexers, tristate drivers, and so forth cannot always be applied via the edge pins, because there may not be enough of them. To overcome this problem, a "test state register" may be incorporated in the design. This could in fact be a shift register that is loaded and controlled by just a few signals. The various testability hardware in the circuit can then be controlled by the parallel outputs of the shift register.

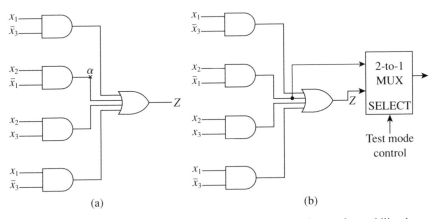

Figure 5.5   (a) Circuit with fault; (b) use of multiplexer to enhance observability during testing

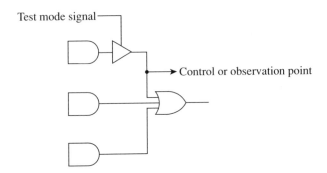

Figure 5.6   Improvement in testability using tristate drivers

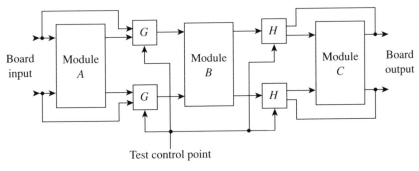

Figure 5.7   Testing of embedded modules (adapted from Ref. 5.1)

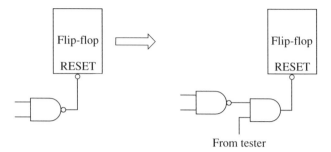

Figure 5.8   Circuit initialization

Frequently, flip-flops, counter, shift registers, and other memory elements assume unpredictable states when power is applied, and they must be set to known
states before testing can begin. Ideally, all memory elements should be reset from
the external pins of the circuit, whereas in some cases additional logic may be
required (Fig. 5.8). With complex circuits it may be desirable to set memory
elements in several known states. This not only allows independent initialization,
it also simplifies generation of certain internal states required to test the board
adequately.

A long counter chain presents another practical test problem. For example, the
counter chain shown in Fig. 5.9 requires thousands of clock pulses to go through
all the states. One way to avoid this problem is to break up the long chain into
smaller ones by using a multiplexer. When the control input c of the multiplexer
is at logic 0, the counter functions normally. When c is at logic 1, the original
counter is partitioned into two smaller counters.

A feedback loop is difficult to test, because it hides the source of a fault. The
source can be located by breaking the loop physically and bringing both lines to
external pins that can be short-circuited for normal operation. When not short-

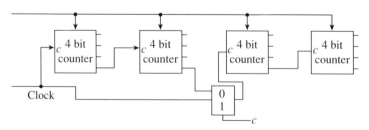

Figure 5.9   Use of a multiplexer to simplify testing of long counter chains

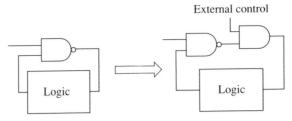

Figure 5.10 Breaking of feedback loop using an extra gate

circuited, the separated lines provide a control point and a test point. An alternative way of breaking a feedback loop, rather than using more costly test/control points, is to add to the feedback path a gate that can be interrupted by a signal from the tester (Fig. 5.10).

5.3 Design of Diagnosable Sequential Circuits

The use of checking experiments to determine whether a sequential circuit represents the behavior specified by its state table yields good results provided that

1. The circuit is reduced and strongly connected.
2. The circuit has a distinguishing sequence.
3. The actual circuit has no more states than the correctly operating circuit.

For circuits that do not have any distinguishing sequences, the checking experiments are very long and consequently hard to apply in any practical situation. One approach to this problem is to modify a given circuit by adding extra outputs so that the modified circuit has a distinguishing sequence. A sequential circuit which possesses one or more distinguishing sequences is said to be *diagnosable*.

A procedure for modifying a sequential circuit to possess a distinguishing sequence if it does not already do so has been presented by Kohavi and Lavelle [5.3]. Let us explain the procedure by considering the state table of Circuit M shown in Fig. 5.11(a); Circuit M does not have a distinguishing sequence. The procedure begins with the construction of the testing table of the circuit; the testing table for Circuit M is shown in Fig. 5.11(b). The column headings consist of all input/output combinations, where the pair X/Z corresponds to input X and output Z. The entries of the table are the "next state." For example, from state A under input 1 the circuit goes to state B with an output of 0. This is denoted

Present state	0/0	0/1	1/0	1/1
A	A	–	B	–
B	A	–	C	–
C	–	A	D	–
D	–	A	A	–
AB	(AA)	–	BC	–
AC	–	–	BD	–
AD	–	–	AB	–
BC	–	–	CD	–
BD	–	–	AC	–
CD	–	(AA)	AD	–

Present state	Next state/output	
	x = 0	x = 1
A	A,0	B,0
B	A,0	C,0
C	A,1	D,0
D	A,1	A,0

(a)   (b)

Figure 5.11 (a) State table of Circuit M; (b) testing table for Circuit M

by entering B in column 1/0 and a dash (–) in column 1/1. In a similar manner, the next states of A are entered in the upper half of the table.

The lower half of the table is derived in a straightforward manner from the upper half. If the entries in rows S_i and S_j, column X_k/Z_1 of the upper half are S_p and S_q respectively, the entry in row S_iS_j, column X_k/Z_1, of the lower half, is S_pS_q. For example, because the entries in rows A and B, column 1/0, B and C, respectively, the corresponding entry in row AB, column 1/0, is BC and so on. If for a pair S_i and S_j either one or both corresponding entries in some column X_k/Z_1 are dashes, the corresponding entry in row S_iS_j, column X_k/Z_1, is a dash. For example, the entry in row BD, column 0/0, is a dash, because the entry in row D, column 0/0, is a dash. Whenever an entry in the testing table consists of a repeated state (e.g., AA in row AB), that entry is circled. A circle around AA implies that both states A and B are merged under input 0 into state A and hence are indistinguishable by any experiment starting with an input 0.

The next step of the procedure is to form the *testing graph* of the circuit. The testing graph is a directed graph with each node corresponding to a row in the lower half of the testing table. A directed edge labeled X_k/Z_1 is drawn from node S_iS_j to node S_pS_q, where $p \neq q$, if there exists an entry in row S_iS_j, column X_k/Z_1 of the testing table. Figure 5.12 shows the testing graph for Circuit M.

A circuit is *definitely diagnosable* if and only if its testing graph has no loops and there are no repeated states, that is, no circled entries in its testing table. Circuit M is therefore not definitely diagnosable, because AA exists in its testing table and its testing graph contains tow loops: $AB–BC–CD–AD–AB$ and $AC–BD–$

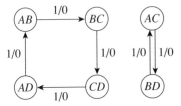

Figure 5.12   Testing graph for Circuit M

AC. To make the circuit definitely diagnosable, additional output variables are required to eliminate all repeated entries from its testing table and to open all loops in its testing graph. The maximum number of extra output terminals required to make a 2^k state circuit definitely diagnosable is k; however, the addition of one output terminal is sufficient to make Circuit M definitely diagnosable. The modified state table of Circuit M is shown in Fig. 5.13; this version possesses the distinguishing sequences 0 and 11. The checking experiment for a definitely diagnosable circuit can be derived as follows:

1. Apply a homing sequence, followed by a transfer sequence (S_i, S_0) is necessary, to bring the circuit into an initial state S_0.

2. Choose a distinguishing sequence so that it is the shorter one of the sequences of all 0s or all 1s. (For the purpose of clearer presentation of the procedure, assume that the distinguishing sequence has been chosen as the all-1s sequence.)

3. Apply the distinguishing sequence followed by a 1. (If the all-0s sequence has been chosen, apply a 0 instead of a 1.)

4. If S_{01}, that is, the 1-successor of S_0, is different from S_0, apply another 1 to check the transition from S_{01} under a 1 input. Similarly, if $S_{011} \neq S_{01}$ and $S_{011} \neq S_0$, apply another 1. Continue to apply 1 inputs in the same manner as long as new transitions are checked.

Present state	Next state/output Z_1Z_2	
	$X = 0$	$X = 1$
A	*A*,00	*B*,01
B	*A*,01	*C*,00
C	*A*,10	*D*,00
D	*A*,11	*A*,01

Figure 5.13   Circuit M with additional output Z_2

5. When an additional 1 input does not yield any new transition, apply an input of 0 followed by the distinguishing sequence.

6. Apply inputs of 1s as long as new transitions can be checked. Repeat steps 5 and 6 when no new transitions can be checked.

7. When steps 5 and 6 do not yield any new transitions and the circuit, which is in state S_i, is not yet completely checked, apply the transfer sequence $T(S_i, S_k)$, where S_k is a state the transition of which has not been checked, such that $T(S_i, S_k)$ passes through checked transitions only.

8. Repeat the last three steps until all transitions have been checked.

The checking experiment for the definitely diagnosable circuit of Fig. 5.13 has been designed using the foregoing procedure. It required only 23 symbols and is illustrated here:

```
Input      1 1 1 1 1 0 1 1 0 1 1 1 1 1 1 1 0 1 1 0 0 1 1
State      A B C D A B A B C A B C D A B C D A B C A A B C
Output Z₁  0 0 0 0 0 0 0 0 1 0 0 0 0 0 0 0 1 0 0 1 0 0 0
       Z₂  1 0 0 1 1 1 1 0 0 1 0 0 1 1 0 0 1 1 0 0 0 1 0
```

For an n-state, m-input circuit, this procedure gives a bound on the length of checking sequences: approximately mn^3.

5.4 The Scan-Path Technique for Testable Sequential Circuit Design

The testing of sequential circuits is complicated because of the difficulties in setting and checking the states of the memory elements. These problems can be overcome by modifying the design of a general sequential circuit so that it will have the following two properties [5.4]:

1. The circuit can easily be set to any desired internal state.

2. It is easy to find a sequence of input patterns such that the resulting output sequence will indicate the internal state of the circuit. In other words: the circuit has a distinguishing sequence.

The basic idea is to add an extra input c to the memory excitation logic in order to control the mode of a circuit. When $c = 0$, the circuit operates in its normal mode, but when $c = 1$, the circuit enters into a mode in which the elements are connected together to form a shift register. This facility is incorporated by inserting a double-throw switch in each input lead of every memory element. All

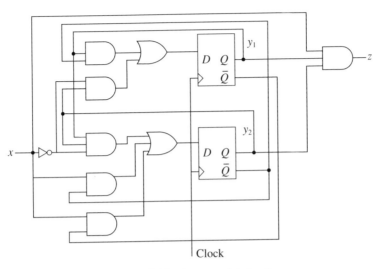

Figure 5.14 A sequential circuit

these switches are grouped together, and the circuit can operate either in its *normal* mode or *shift register* mode. Figure 5.14 shows a sequential circuit using D flip-flops; the circuit is modified as shown in Fig. 5.15. Each of the double-throw switches may be realized as indicated in Fig. 5.16. One additional input

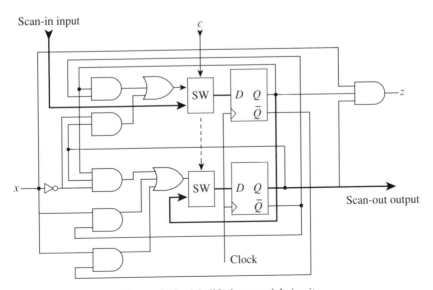

Figure 5.15 Modified sequential circuit

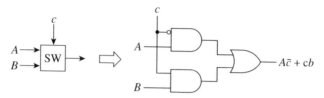

Figure 5.16   A realization for the double-throw switch

connection to the modified circuit is required to supply the signal c to control all the switches.

In the shift register mode, the first flip-flop can be set directly from the primary inputs (*scan-in* inputs) and the output of the last flip-flop can be directly monitored on the primary output (*scan-out* output). This means that the circuit can be set to any desired state via the scan-in inputs, and that the internal state can be determined via the scan-out output. The procedure for testing the circuit is as follows:

1. Set $c = 1$ to switch the circuit to shift register mode.
2. Check operation as a shift register by using scan-in inputs, scan-out output, and the clock.
3. Set the initial state of the shift register.
4. Set $c = 0$ to return to normal mode.
5. Apply test input pattern to the combinational logic.
6. Set $c = 1$ to return to shift register mode.
7. Shift out the final state while setting the starting state for the next test.
8. Go to step 4.

With this procedure a considerable proportion of the actual testing time is spent in setting the state, an operation that requires a number of clock pulses equal to the length of the shift register. This time may be decreased by forming several short shift registers rather than a single long one; the time needed to set or read the state would then be equal to the length of the longest shift register. The extent to which the number of shift registers can be increased is determined by the number of input and output connections available to be used to drive and sense the shift registers.

The main advantage of the scan-path approach is that a sequential circuit can be transformed into a combinational circuit, thus making test generation for the circuit relatively easy. Besides, very few extra gates or pins are required for this transformation.

Another implementation of the scan-path technique has been described by Funatsu *et al.* [5.5]. The basic memory element used in this approach is known as a *raceless D-type flip-flop with scan path* [5.6] Figure 5.17(a) shows such a memory element, which consists of two latches *L1* and *L2*. The two clock signals *C1* and *C2* operate exclusively. During normal operation, *C2* remains at logic 1 and *C1* is set to logic 0 for sufficient time to latch up the data at the data input *D1*. The output of *L1* is latched into *L2* when *C1* returns to logic 1.

Scan-in operation is realized by clocking the test input value at *D2* into the latch *L1* by setting *C2* to logic 0. The output of the *L1* latch is clocked into *L2* when *C2* returns to logic 1.

The configuration of the scan-path approach used at logic card level is shown in Fig. 5.17(b). All the flip-flops on a logic card are connected as a shift register, such that for each card there is one scan path. In addition, there is provision for selecting a specified card in a subsystem with many cards by *X–Y* address signals (Fig. 5.17(b)). If a card is not selected, its output is blocked; thus, a number of card outputs in a subsystem can be put together with only a particular card having control of the test output for that subsystem. The Nippon Electric Company in Japan has adopted this version of the scan path approach to improve the testability of their FLT-700 processor system.

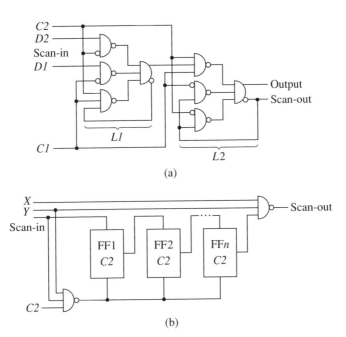

(a)

(b)

Figure 5.17   (a) Raceless flip-flop with scan path; (b) configuration of logic card (adapted from Ref. 5.5)

5.5   Level-Sensitive Scan Design (LSSD)

One of the best known and the most widely practiced methods for synthesizing testable sequential circuits is the IBM LSSD (level-sensitive scan design) [5.7–5.10]. The *level-sensitive* aspect of the method means that a sequential network is designed so that the steady-state response to any input state change is independent of the component and wire delays within the network. Also, if an input state change involves the changing of more than one input signal, the response must be independent of the order in which they change. These conditions are ensured by the enforcement of certain design rules, particularly pertaining to the clocks that evoke state changes in the network. *Scan* refers to the ability to shift into or out of any state of the network.

5.5.1   CLOCKED HAZARD-FREE LATCHES

In LSSD, all internal storage is implemented in hazard-free *polarity-hold* latches. The polarity-hold latch has two input signals as shown in Fig. 5.18(a). The latch cannot change state if $C = 0$. If C is set to 1, the internal state of the latch takes the value of the excitation input D. A flow table for this sequential network, along with an excitation table and a logic implementation, is shown in Fig. 5.18.

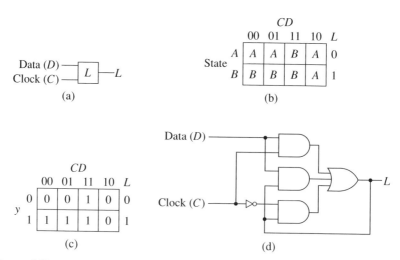

Figure 5.18   Hazard-free polarity-hold latch: (a) symbolic representation; (b) flow table; (c) excitation table; (d) logic implementation (from E. B. Eichelberger and T. W. Williams, ''A logic design structure for LSI testability,'' Proc. 14th Design Automation Conf., June 1978. Copyright © 1978 IEEE. Reprinted with permission).

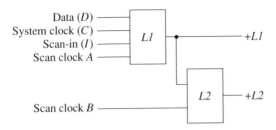

Figure 5.19 Polarity-hold shift register latch (from E. B. Eichelberger and T. W. Williams, ''A logic design structure for LSI testability,'' Proc. 14th Design Automation Conf., June 1978. Copyright © 1978 IEEE. Reprinted with permission).

The clock signal C will normally occur (change from 0 to 1) after the data signal D has become stable at either a 1 or a 0. The output of the latch is set to the new value of the data signal at the time the clock signal occurs. The correct changing of the latch does not depend on the rise or fall time of the clock signal, only on the clock signal being 1 for a period equal to or greater than the time required for the data signal to propagate through the latch and stabilize.

A shift register latch (SRL) can be formed by adding a clocked input to the polarity-hold latch $L1$ and including a second latch $L2$ to act as intermediate storage during shifting (Fig. 5.19). As long as the clock signals A and B are both 0, the $L1$ latch operates exactly like a polarity-hold latch. Terminal I is the scan-in input for the shift register latch and $+L2$ is the output. The logic implementation of the SRL is shown in Fig. 5.20. When the latch is operating as a shift

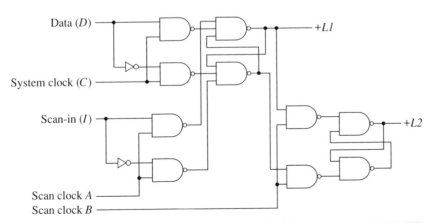

Figure 5.20 Logic for the shift-register latch (from E. B. Eichelberger and T. W. Williams, ''A logic design structure for LSI testability,'' Proc. 14th Design Automation Conf., June 1978. Copyright © 1978 IEEE. Reprinted with permission).

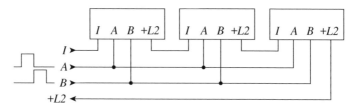

Figure 5.21 Linkage of several SRLs (from E. B. Eichelberger and T. W. Williams, "A logic design structure for LSI testability," Proc. 14th Design Automation Conf., June 1978. Copyright © 1978 IEEE. Reprinted with permission).

register, data from the preceding stage are gated into the polarity-hold switch via I, through a change of the clock A from 0 to 1. After A has changed back to 0, clock B gates the data in the latch $L1$ into the output latch $L2$ Clearly, A and B can never both be 1 at the same time if the shift register latch is to operate properly.

The SRLs can be interconnected to form a shift register as shown in Fig. 5.21. The input I and the output $+L2$ are strung together in a loop, and the clocks A and B are connected in parallel.

5.5.2   LSSD DESIGN RULES

A specific set of design rules has been defined to provide level-sensitive logic subsystems with a scannable design that would aid testing [5.8].

Rule 1   All internal storage is implemented in hazard-free polarity-hold latches.

Rule 2   The latches are controlled by two or more nonoverlapping clocks such that

(a) Two latches, where one feeds the other, cannot have the same clock (see Fig. 5.22(a)).

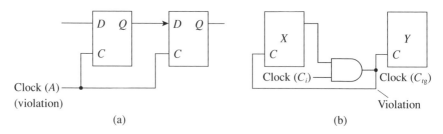

Figure 5.22   (a) Checking Rule 2(a); (b) checking Rule 2(b)

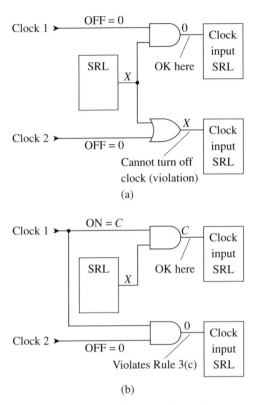

Figure 5.23   (a) Clock OFF test. (b) clock ON test (from Funatsu, *et al.,* "Test generation systems in Japan," Proc. 12th Design Automation Conf., 1975. Copyright © 1978 IEEE. Reprinted with permission).

(b) A latch X may gate a clock C_i to produce a gated clock C_{ig}, which drives another latch Y if and only if clock C_{ig} does not clock latch X, where C_{ig} is any clock derived from C_i (see Fig. 5.22(b)).

Rule 3   It must be possible to identify a set of clock primary inputs from which the clock inputs to SRLs are controlled either through simple powering trees or through logic that is gated by SRLs and/or nonclock primary inputs (see Fig. 5.23). Given this structure, the following rules must hold:

(a) All clock inputs to all SRLs must be at their OFF states when all clock primary inputs are held to their OFF states (see Fig. 5.23(a)).

(b) The clock signal that appears at any clock input of any SRL must be controllable from one or more clock primary inputs, such that it is possible to set the clock input of the SRL to an ON state by turning any one

of the corresponding clock primary inputs to its ON state and also setting the required gating condition from SRLs and/or nonclock primary
inputs.

(c) No clock can be ANDed with either the true value or the complement
value of another clock (see Fig. 5.23(b)).

Rule 4   Clock primary inputs may not feed the data inputs to latches either
directly or through combinational logic, but they may only feed the clock input
to the latches or the primary outputs.

A sequential logic network designed in accordance with Rules 1–4 would be
level-sensitive. To simplify testing and minimize the primary inputs and outputs,
it must also be possible to shift data into and out of the latches in the system.
Therefore, two more rules must be observed:

Rule 5   All SRLs must be interconnected into one or more shift registers, each
of which has an input, an output, and clocks available at the terminals of the
module.

Rule 6   There must exist some primary input sensitizing condition (referred to
as the scan state) such that

(a) Each SRL or scan-out primary output is a function of only the preceding
SRL or scan-in primary input in its shift register during the shifting operation.

(b) All clocks except the shift clocks are held OFF at the SRL inputs.

(c) Any shift clock to an SRL may be turned ON and OFF by changing the
corresponding clock primary input for each clock.

A sequential logic network that is level-sensitive and also has the scan
capability as per Rules 1 to 6 is called a level-sensitive scan design (LSSD).
Figure 5.24(a) depicts a general structure for an LSSD system in which all system
outputs are taken from the $L2$ latch; hence, it is called a *double-latch design*.
In the double-latch configuration, each SRL operates in a master–slave mode.
Data transfer occurs under system clock and scan clock B during normal operation, and under scan clock A and scan clock B during scan-path operation. Both
latches are therefore required during system operation. In the *single-latch* con-
figuration, the combinational logic is partitioned into two disjoint sets, Comb1
and Comb2 (Fig. 5.24(b)). The system clocks used for SRLs in Comb1 and
Comb2 are denoted by Clock1 and Clock2, respectively; they are nonoverlapping.
The outputs of the SRLs in Comb1 are fed back as secondary variable inputs to
Comb2, and vice versa. This configuration uses the output of latch $L1$ as the

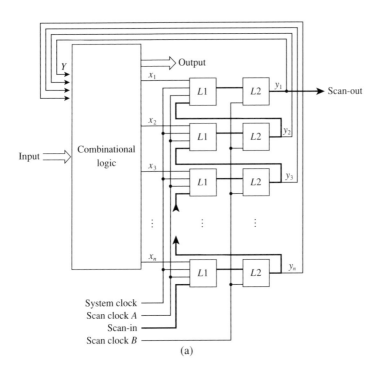

(a)

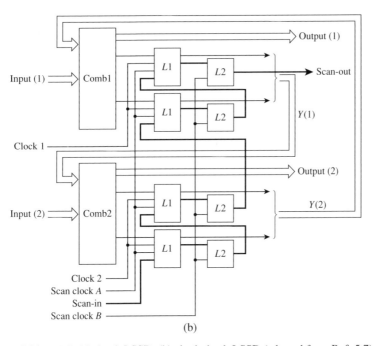

(b)

Figure 5.24 (a) Doble-latch LSSD; (b) single-latch LSSD (adapted from Ref. 5.7)

119

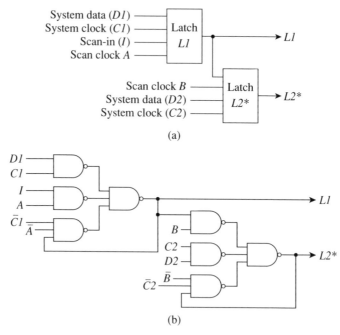

Figure 5.25 (a) $L1/L2^*$ SRL; (b) implementation of $L1/L2^*$ SRL using NAND gates (adapted from Ref. 5.9)

system output; the $L2$ latch is used only for shifting. In other words, the $L2$ latches are redundant and represent an overhead for testability. However, the basic SRL design can be modified to reduce the overhead. The modified latch called the $L1/L2^*$ SRL is shown in Fig. 5.25. The main difference between the basic SRL and the $L1/L2^*$ SRL is that $L2^*$ has an alternative system data input $D2$ clocked in by a separate system clock $C2$. The original data input D in Fig. 5.19 is also available and is now identified as $D1$ in Fig. 5.25. $D1$ is clocked in by the original system clock, which is now called $C1$. Clock signals $C1$ and $C2$ are nonoverlapping. The single-latch configuration of Fig. 5.25(b) can now be modified to the configuration of Fig. 5.26, in which the system output can be taken from either the $L1$ output or the $L2^*$ output. In other words, both $L1$ and $L2^*$ are utilized, which means fewer latches are required in the system. As a result, there is a significant reduction in the silicon cost when $L1/L2^*$ SRLs are used to implement the LSSD. Although both latches in $L1/L2^*$ SRLs can be used for system functions, it is absolutely essential, as in conventional LSSD, that both $L1$ and $L2^*$ outputs do not feed the same combinational logic.

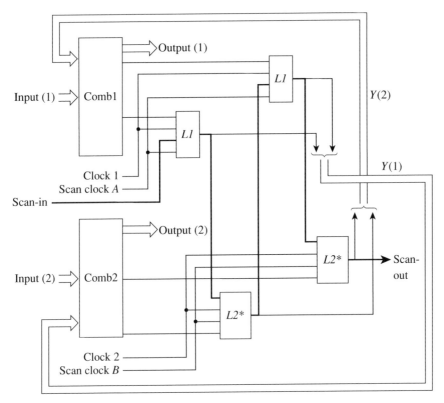

Figure 5.26   Single-latch LSSD design using SRL *L1/L2** (adapted from Ref. 5.9)

5.5.3 ADVANTAGES OF THE LSSD TECHNIQUE

The LSSD approach is very similar to the scan-path approach used by the NEC, except that it has the level-sensitive attribute and requires two separate clocks to operate latches *L1* and *L2*. The use of LSSD alleviates the testing problems in the following ways:

1. The correct operation of the logic network is independent of a.c. charac-teristics such as clock edge rise time and fall time.
2. Network is combinational in nature as far as test generation and testing is concerned.
3. The elimination of all hazards and races greatly simplifies both test gen-eration and fault simulation.

The ability to test networks as combinational logic is one of the most important benefits of the LSSD. This is done by operating the polarity-hold latches as SRLs during testing.

Any desired pattern of 1s and 0s can be shifted into the polarity-hold latches as inputs to the combinational network. For example, the combinational network of Fig. 5.24(a) is tested by shifting part of each required pattern into the SRLs, with the remainder applied through the primary inputs. Then the system clock is turned on for one cycle, the test pattern is propagated through the combinational logic, and the result of the test is captured in the register and at the primary outputs. The result of the test captured in the register is then scanned out and compared with the expected response. The shift register must also be tested, and this is accomplished by shifting a short sequence of 1s and 0s through the shift register latches.

In general, most functions designed in an unconstrained environment can be designed using the LSSD approach with little or no impact on performance. However, the requirement of level-sensitive flip-flops demand custom-designed chips; hence, the LSSD approach would be difficult to apply to a board-level circuit built using off-the-shelf ICs.

5.6   Random Access Scan Technique

The design methods discussed in Secs. 5.4 and 5.5 use sequential access scan-in/scan-out techniques to improve testability; that is, all flip-flops are connected in series during testing to form a shift register or registers. In an alternative approach, known as *random access scan,* each flip-flop in a logic network is selected individually by an address for control and observation of its state [5.11]. The basic memory element in a random access scan-in/scan-out network is an *addressable latch.* The circuit diagram of an addressable latch is shown in Fig. 5.27. A latch is selected by X–Y address signals, the state of which can then be controlled and observed through scan-in/scan-out lines. When a latch is selected and its scan clock goes from 0 to 1, the scan data input is transferred through the network to the scan data output, where the inverted value of the scan data can be observed. The input on the DATA line is transferred to the latch output Q during the negative transition (1 to 0) of the clock. The scan data out lines from all latches are ANDed together to produce the chip scan-out signal; the scan-out line of a latch remains at logic 1 unless the latch is selected by the X–Y signals.

A different type of addressable latch—the set/reset type—is shown in Fig. 5.28. The "clear" signal clears the latch during its negative transition. Prior

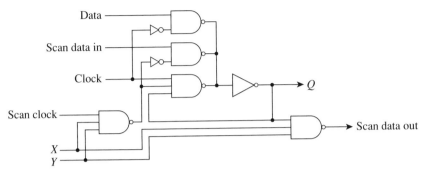

Figure 5.27 An addressable latch (from H. Ando, "Testing of VLSI with random access scan," Proc. COMPCON Spring 1980. Copyright © 1980 IEEE. Reprinted with permission).

to scan-in operation, all latches are cleared. Then, a latch is addressed by the $X–Y$ lines and the preset signal is applied to set the latch state.

The basic model of a sequential circuit with random access scan-in/scan-out network is shown in Fig. 5.29. The X- and Y-address decoders are used to access an addressable latch—like a cell in random access memory. A tree of AND gates is used to combine all scan-out signals. Clear input of all latches are tied together to form a master reset signal. Preset inputs of all latches receive the same scan-in signal gated by the scan clock; however, only the latch accessed by the $X–Y$ addresses is affected.

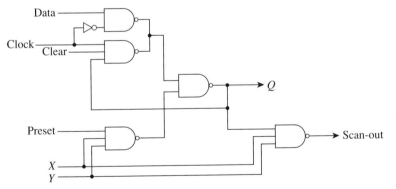

Figure 5.28 Set/reset type addressable latch (from H. Ando, "Testing of VLSI with random access scan," Proc. COMPCOM Spring 1980. Copyright © 1980 IEEE. Reprinted with permission).

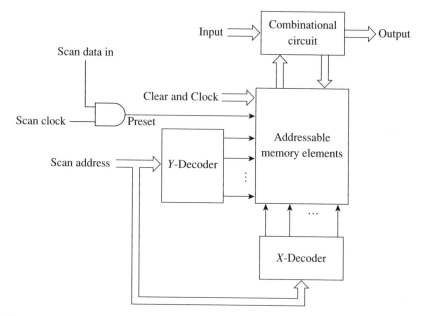

Figure 5.29 Sequential circuit designed with addressable latches (from H. Ando, "Testing of VLSI with random access scan," Proc. COMPCON Spring 1980. Copyright © 1980 IEEE. Reprinted with permission).

The test procedure of a network with random access scan-in/scan-out network is as follows:

1. Set test input to all test points.
2. Apply the master reset signal to initialize all memory elements.
3. Set scan-in address and data, and then apply the scan clock.
4. Repeat step 3 until all internal test inputs are scanned in.
5. Clock once for normal operation.
6. Check states of the output points.
7. Read the scan-out states of all memory elements by applying appropriate X–Y signals.

The random access scan-in/scan-out technique has several advantages:

1. The observability and controllability of all system latches are allowed.
2. Any point in a combinational network can be observed with one additional gate and one address per observation point.

3. A memory array in a logic network can be tested through a scan-in/scan-out network. The scan address inputs are applied directly to the memory array. The data input and the write-enable input of the array receive the scan data and the scan clock, respectively. The output of the memory array is ANDed into the scan-out tree to be observed.

The technique has also a few disadvantages:

1. Extra logic in the form of two address gates for each memory element, plus the address decoders and output AND trees, result in 3–4 gates overhead per memory element.
2. Scan control, data, and address pins add up to 10–20 extra pins. By using a serially loadable address counter, the number of pins can be reduced to around 6.
3. Some constraints are imposed on the logic design, such as the exclusion of asynchronous latch operation.

5.7 Partial Scan

In full scan, all flip-flops in a circuit are connected into one or more shift registers; thus, the states of a circuit can be controlled and observed via the primary inputs and outputs, respectively. In partial scan, only a subset of the circuit flip-flops are included in the scan chain in order to reduce the overhead associated with full scan design [5.12]. Figure 5.30 shows a structure of partial scan design. This structure has two separate clocks: a system clock and a scan clock. The scan clock controls only the scan flip-flops. Note that the scan clock is derived by gating the system clock with the scan-enable signal; no external clock is necessary. During the normal mode of operation, namely, scan-enable signal at logic 0, both scan and nonscan flip-flops update their states when the system clock is applied. In the scan mode operation, only the state of the shift register (constructed from the scan flip-flops) is shifted one bit with the application of the scan flip-flop; the nonscan flip-flops do not change their states.

The disadvantage of two-clock partial scan is that the routing of two separate clocks with small skews is very difficult to achieve. Also, the use of a separate scan clock does not allow the testing of the circuit at its normal operating speed. Cheng [5.13] proposed a partial scan scheme, shown in Fig. 5.31, in which the system clock is also used as the scan clock. Both scan and nonscan flip-flops

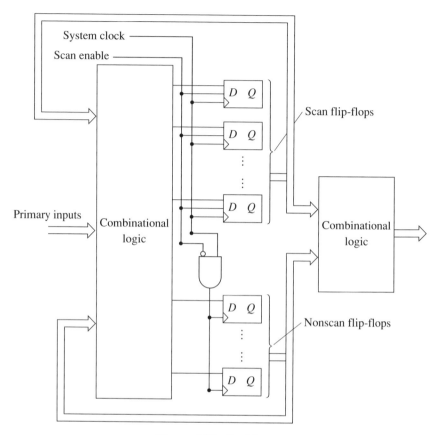

Figure 5.30 Partial scan

move to their next states when the system clock is applied. A test sequence is derived by shifting data into the scan flip-flops. This data together with the contents of nonscan flip-flops constitute the starting state of the test sequence. The other patterns in the sequence are obtained by single-bit shifting of the contents of scan flip-flops, which form part of the required circuit states. The remaining bits of the states, that is, the contents of nonscan flip-flops, are determined by the functional logic. Note that this form of partial scan scheme allows only a limited number of valid next states to be reached from the starting state of the test sequence. This may limit the fault coverage obtained by using the technique.

The selection flip-flops to be included in the partial scan is done by heuristic methods. Trischler [5.14] has used testability analysis to show that the fault coverage in a circuit can be significantly increased by including 15–25% of the flip-

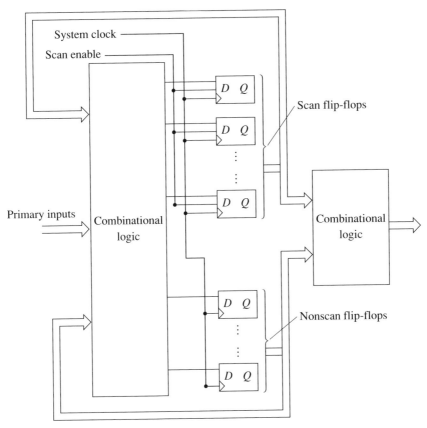

Figure 5.31   Partial scan using the system clock

flops in the partial scan. Agrawal *et al.* [5.15] have shown that the fault coverage can be increased to as high as 95% by including less than 65% of the flip-flops in the partial scan.

5.8   Testable Sequential Circuit Design Using Nonscan Techniques

Full and partial scan techniques improve the controllability and observability of flip-flops in sequential circuit, and therefore the test generation for such circuits is considerably simplified. However, a scan-based circuit cannot be tested at its normal speed, because test data have to be shifted in and out via the scan path.

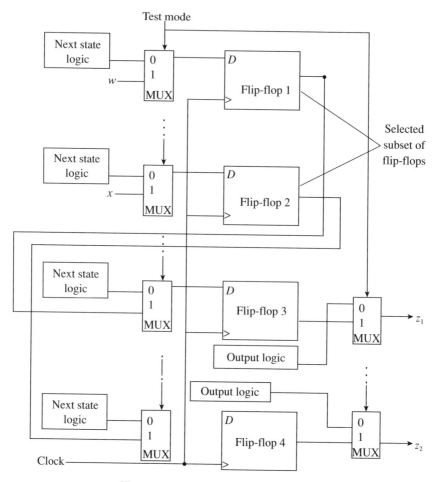

Figure 5.32   Testable sequential circuit

Reddy and Dandapani [5.16] have developed a strategy for implementing sequential circuits that is not based on scan flip-flops. The original sequential circuit is augmented with a *test mode* input. If this input is at logic 1, each flip-flop from a selected subset of flip-flops in the circuit is connected directly to a primary input. This enables data to be shifted directly into these flip-flops from the primary inputs; at the same time, previous data from these flip-flops are shifted into the flip-flops that do not belong to the subset. The contents of the flip-flops in the selected subset are available on the primary outputs when the test mode signal is at logic 1. To illustrate let us consider the two-input and two-output sequential circuit with four flip-flops shown in Fig. 5.32. The primary inputs w and x are used to shift data into selected flip-flops 1 and 2, respectively, when the *test mode* signal is at logic 1. The previous contents of flip-flops 1 and 2 are shifted into flip-flops 3 and 4, respectively, while the contents of flip-flops 3 and 4 are observable via primary outputs z_1 and z_2.

Chickermane *et al.* [5.17] have proposed several techniques similar to that in Ref. 5.16 to enhance controllability and observability in sequential circuits so that the testing of the circuits can be done at normal speed. The controllability of a circuit is improved by selecting a subset of flip-flops such that the number of cycles in the circuit's state diagram is minimized and each selected flip-flop can be loaded directly from one primary input line during the test mode. Such flip-flops are identified as *controllable* flip-flops. Figure 5.33 shows a sequential circuit modified for enhanced controllability. When the *test* input is at logic 0, all the flip-flops in a circuit are driven by their original next state logic. In other words, the circuit operates in its normal mode. When the *test* input is at logic 1, each controllable flip-flop is driven by a primary input line via a 2-to-1 multiplexer. The observability is improved by selecting a set of internal nodes that are untestable because faults at such nodes cannot be propagated to the primary outputs. The signals at these nodes are compressed by an EX–OR tree, the output of which is available on an additional output line as shown in Fig. 5.34.

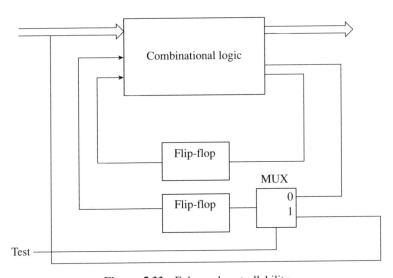

Figure 5.33   Enhanced controllability

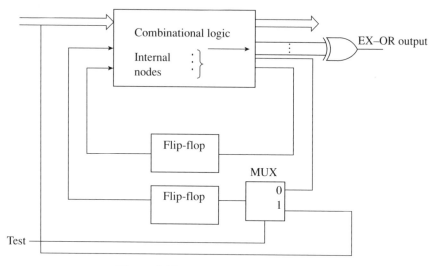

Figure 5.34   Observability enhancement

5.9   CrossCheck

The CrossCheck approach incorporates test circuitry into the basic cells used to implement VLSI design [5.18,5.19]. This is done by connecting the output of a cell to the drain of a pass transistor. The gate of the transistor is connected to a probe line P and the source to a sense line S, as shown in Fig. 5.35. The output of the cell can be observed via the S-line by controlling the probe line P. In other words, the controllability and the observability of the cells can be guaranteed.

    This approach can be used to enhance testability of VLSI chips by using cells with CrossCheck test points to implement logic. The individual test points are routed into an orthogonal grid of probe and sense lines as shown in Fig. 5.36.

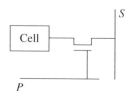

Figure 5.35   Basic cell

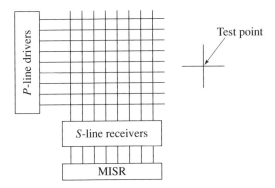

Figure 5.36 CrossCheck grid

Appropriate test patterns are first applied to the circuit under test via the primary inputs. The probe lines are then enabled, allowing logic values at internal test points to be transferred to the sense lines. These values are stored in a parallel-in/serial-out shift register. Alternatively, an MISR (Multiple Input Signature Register) can be used to compress the test values into a signature. It should be clear from this discussion that the CrossCheck approach allows parallel scanning of test points.

The CrossCheck testability approach simplifies the detection of stuck-open faults in logic cells. To illustrate, let us consider a two-input NOR gate augmented by a transistor having a *P*-line and an *S*-line (Fig. 5.37). Let us assume the NOR

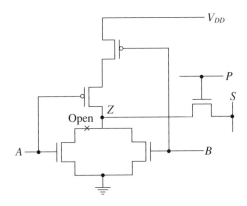

Figure 5.37 A two-input CMOS NOR gate with a stuck-open fault

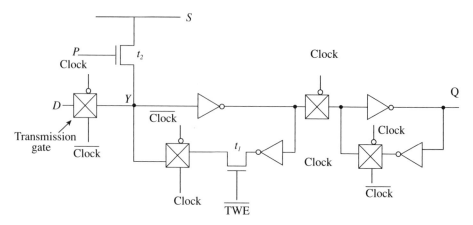

Figure 5.38   Cross-controlled latch

gate has a stuck-open fault as shown in Fig. 5.37. The fault can be activated by applying the test pattern $AB = 10$. The sense line is precharged to a *weak* logic 1. When the probe line P is set at logic 1, the logic value on the S-line will be 0 in the absence of the assumed stuck-open fault; otherwise, the logic value will be 1. Thus, a stuck-open fault can be detected with a single test pattern.

The CrossCheck approach enhances the controllability of sequential circuits by using a D flip-flop implementation called *cross-controlled latch (CCL)*. A CCL consists of a master–slave edge-triggered flip-flop augmented by two transistors t_1 and t_2 (Fig. 5.38). Transistor t_1 is controlled by a test–write–enable ($\overline{\text{TWE}}$) signal. When $\overline{\text{TWE}} = 1$, and also probe line P and the clock are set at logic 1, the value on node Y is observable on the sense line S via transistor t_2; this value is also observable at the Q output. When the $\overline{\text{TWE}}$ signal is set at logic 0, transistor t_1 is turned off and the feedback path is disabled. A CCL functions as a D flip-flop when the $\overline{\text{TWE}}$ signal is at logic 1. The overhead due to the extra transistors is around 3%; also, there is a slight increase in the minimum clock pulse width. On the plus side, the setting of flip-flops to specific values can be done in parallel; thus, the test generation process for sequential circuits may be faster than in scan-based circuits.

The CrossCheck approach allows high observability of test points in a circuit. This, in conjunction with the enhanced controllability resulting from the use of CCLs, makes the CrossCheck approach a powerful design for testability technique. The drawback of the approach is that the scanning of test points introduces additional delay in the test application time. Thus, as in the scan technique, it is not possible to perform testing at the operational speed of the circuit.

5.10   Boundary Scan

Boundary scan methodology is used to resolve the problem of controlling and observing the input and output pins of chips used in assembling a system/sub-system on a printed circuit board (PCB). An international consortium, the Joint Test Action Group (JTAG), proposed a boundary scan architecture that was adopted by the IEEE Standards Board as IEEE Std. 1149.1. This architecture provides a single serial scan path through the input/output pins of individual chips on a board. The scan path is created by connecting the normal inputs/ouputs of the chip logic to the input/output pins of the chip through *boundary scan cells*. Figure 5.39 shows an implementation of a boundary scan cell. The operation of the boundary scan cells is controlled by the Test Access Port (TAP), which has four inputs: TCK (Test Clock), TMS (Test Mode Select), TDI (Test Data Input), and TDO (Test Data Output). A chip with boundary scan cells placed next to the pins is shown in Fig. 5.40(a). These cells can be interconnected to form a continuous shift register (the boundary scan path) around the border of the chip. This shift register, as shown in Fig. 5.40(b), can be used to shift, apply, or capture test data. During normal operation of a chip, the boundary scan cells are transparent, and data applied at the input pins flows directly to the circuit. The corresponding output response of the circuit flows directly to the output pins of the chip. During the scan mode test patterns are shifted in via the TDI pin, and the captured responses are shifted out via the TDO pin. Thus, the controllability and observability of the pins of a chip can be achieved without making the pins physically accessible.

In addition to the TAP, the boundary scan architecture includes a controller

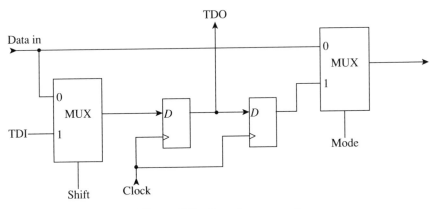

Figure 5.39   Boundary scan cell

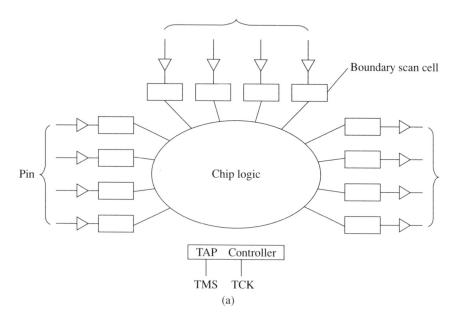

(a)

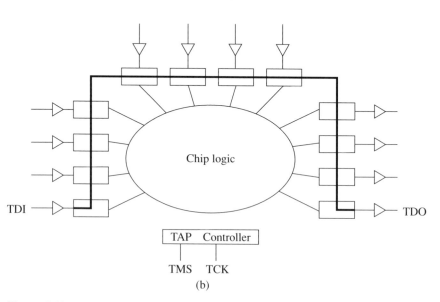

(b)

Figure 5.40   (a) Chip with boundary scan cells placed next to the pins; (b) scan cells interconnected to form boundary scan register

for the TAP, an instruction register (IR), and a group of test data registers (TDR). Figure 5.41 shows a JTAG boundary scan architecture. The TAP controller is a finite state machine that generates various control signals—update, shift, or capture data—required in the boundary scan architecture. The state transitions in the TAP controller occur on the rising edge of the clock pulse at the TCK pin. The instruction register is a serial-in/serial-out shift register; the contents of the shift register are stored in a parallel output latch. Once the contents of the shift register are stored in a parallel output latch. Once the contents of the shift register, i.e.,

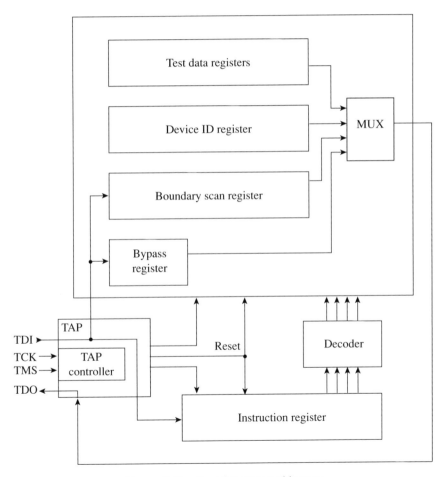

Figure 5.41   Boundary scan architecture

the current instruction is loaded into the output latch, it can only be changed when the TAP controller is in either update–IR or Test–Logic–Reset state. In the update–IR state, the newly shifted instruction is loaded into the output latch, whereas in the Test–Logic–Reset state, all test logic is reset and the chip performs its normal operation. The TAP controller can be put in the Test–Logic–Reset state from any other state by holding the TMS signal at logic 1 and applying the TCK pulse at least five times.

The boundary scan architecture, in addition to the boundary scan register, contains another data register, called the bypass register. The bypass register is only 1 bit wide. As can be seen in Fig. 5.41, data at the TDI input of a chip can be moved to its TDO output through the bypass register in one clock cycle, rather than the multiple clock cycles required to shift data along the length of the scan path. Thus, the bypass register is extremely useful if only a small subset of all the chips on a board need to be tested. Optionally, a device identification register may be included in a chip together with the test data registers. It is a 32-bit parallel-in/serial-out shift register and is loaded with data that identify the chip manufacturer's name, part number, and version number when the TAP controller is in Capture–DR (Data Register) state. The normal operation of a chip is not affected if either the ID register or the bypass register is in use.

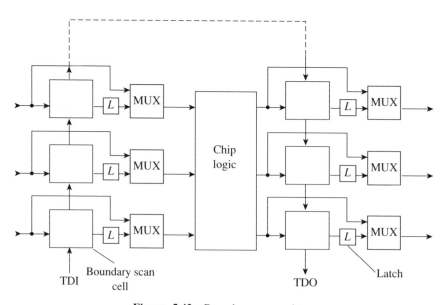

Figure 5.42 Boundary scan register

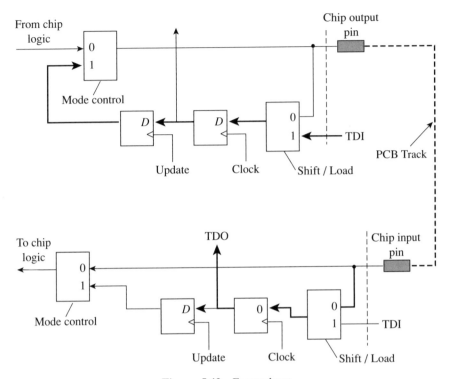

Figure 5.43   External test

As mentioned previously, boundary scan cells can be interconnected to form a shift register. This shift register is identified as the *boundary scan register*; the structure of this register is shown in Fig. 5.42. The outputs of all shift register stages are stored in a latch in order to prevent the change of information while data is being shifted in and out.

A boundary scan register can be configured to perform three types of tests: *external test, internal test,* and *sample test.* The external test is used to test interconnections for stuck-at and bridging faults. It is invoked by entering logic 0 into every cell of the instruction register. Test patterns are then shifted into the boundary scan register stages at the output pins of a chip via the TDI pin. These patterns arriving at the input pins of other chips are loaded into their boundary scan registers, and they are observed via the TDO pins (Fig. 5.43).

The internal test allows individual chips on a PCB to be tested. During this mode, test vectors are shifted into the boundary scan register of the chip under test via the TDI pin. The corresponding test responses are loaded into the boundary scan register, and shifted out for verification. Thus, the internal test allows individual chips on a board to be tested without using a bed-of-nails interface.

The sample test allows monitoring of data flowing into and out of a chip during its normal operation. This can aid an external tester in understanding the chip's performance.

References

5.1   Muehldorf, E. I., "Designing LSI logic for testability," *Proc. Semicond. Test Conf.*, 45–49 (1976).
5.2   Oberly, R. P., "How to beat the card test game," *Proc. Semicond. Test Conf.*, 16–18 (1977).
5.3   Kohavi, Z., and P. M. Lavelle, "Design of sequential machines with fault detection capabilities," *IEEE Trans. Comput.*, 473–484 (August 1967).
5.4   Williams M. J. Y., and J. M. Angell, "Enhancing testability of large-scale integrated circuits via test points and additional logic," *IEEE Trans. Comput.*, 46–59 (January 1973).
5.5   Funatsu, S. N., N. Wakatsuki, and T. Armia, "Test generation systems in Japan," *Proc. ACM/IEEE Design Automation Conf.*, 114–122 (1975).
5.6   Yamada, A., *et al.*, "Automatic system level test generation and fault location for large digital systems," *Proc. ACM/IEEE Design Automation Conf.*, 347–352 (1978).
5.7   Eichelberger, E. B., and T. W. Williams, "A logic system structure for LSI testability," *Proc. ACM/IEEE Design Automation Conf.*, 462–468 (1978).
5.8   Godoy, H. C., G. B. Franklin, and P. S. Bottorff, "Automatic checking of logic design structure for compliance with testability ground rules," *Proc. ACM/IEEE Design Automation Conf.*, 469–478 (1978).
5.9   Dasgupta, S., P. Goel, R. G. Walther and T. W. Williams, "A variation of LSSD and its implications on design and test pattern generation in VLSI," *Proc. Intl. Test Conf.*, 63–66 (1982).
5.10   Williams, T. W., and K. P. Parker, "Design for testability—a survey," *IEEE Trans. Comput.*, 2–15 (January 1982).
5.11   Ando, H., "Testing of VLSI with random access scan," *Proc. COMPCON* 50–52 (Spring 1980).
5.12   Cheng, K. T., and V. D. Agarwal, "A partial scan method for sequential circuits with feedback," *IEEE Trans. Comput.*, 544–548 (April 1990).
5.13   Cheng, K. T., "Single clock partial scan," *IEEE Design and Test of Computers*, 24–31 (Summer 1995).
5.14   Trischler, E., "Testability analysis and complete scan path," *Proc. Intl. Conf. on CAD*, 38–39 (1983).

5.15 Agarwal, V. D., K. T. Cheng, D. D. Johnson, and T. Lin, "A complete solution to the partial scan problem," *Proc. Intl. Test Conf.*, 41–51 (1987).

5.16 Reddy, S. M., and R. Dandapani, "Scan design using standard flip-flops," *IEEE Design and Test of Computers*, 52–54 (February 1987).

5.17 Chickermane, V., E. M. Rudnick, P. Banerjee, and J. H. Patel, "Non-scan design-for-testability techniques for sequential circuits," *Proc. ACM/IEEE Design Automation Conf.*, 236–241 (1993).

5.18 Gheewala, T., "CrossCheck: A cell based VLSI testability solution," *Proc. ACM/IEEE Design Automation Conf.*, 706–709 (1989).

5.19 Chandra, J. C., T. Ferry, T. Gheewala, and K. Pierce, "ATPG based on a novel grid-addressable latch element," *Proc. ACM/IEEE Design Automation Conf.*, 282–286 (1991).

5.20 John Fluke Mfg. Co. Inc. *The ABCs of Boundary-Scan Test* (1991).

Chapter 6 | Built-In Self-Test

Advances in VLSI technology have led to the fabrication of chips that contain a very large number of transistors. The task of testing such a chip to verify correct functionality is extremely complex, and often very time consuming. In addition to the problem of testing the chips themselves, the incorporation of these into systems has caused the cost of test generation to grow exponentially. For example, the approximate cost of detecting a fault at the board level is 10 times as high as detecting a fault at the chip level, and the cost increases by about 10-fold from board level to system level [6.1]. A widely accepted approach to deal with the testing problem at the chip level is to incorporate built-in self-test (BIST) capability inside a chip. This increases the controllability and the observability of the chip, thereby making the test generation and fault detection easier. The conventional testing, test patterns are generated externally by using CAD (computer-aided design) tools. The test patterns and the expected responses of the circuit under test to these test patterns are used by an ATE (Automatic Test Equipment) to determine if the actual responses match the expected ones. On the other hand, in BIST the test pattern generation and the output response evaluation are done on chip; thus, the use of expensive ATE machines to test chips can be avoided. A basic BIST structure is shown in Fig. 6.1. The main function of the test pattern generator is to apply test patterns to the circuit under test (assumed to be a multioutput combinational circuit). The resulting output patterns are transferred to the output response analyzer. Ideally, a BIST scheme should be easy to implement and must provide a high fault coverage.

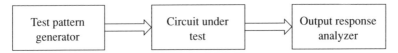

Figure 6.1   BIST structure

6.1   Test Pattern Generation for BIST

Test pattern generation approaches for BIST schemes can be divided into two categories:

Exhaustive/pseudo-exhaustive testing

Pseudo-random testing

Deterministic testing

6.1.1   EXHAUSTIVE TESTING

In the exhaustive testing approach, all possible input patterns are applied to the circuit under test. Thus, for an n-input combinational circuit, all possible 2^n patterns need to be applied. The advantage of this approach is that all irredundant faults can be detected; however, any fault that converts the combinational circuit into a sequential circuit, for example, a bridging fault, cannot be detected. The drawback of this approach is that when n is large, the test application time becomes prohibitive, even with high clock speeds. Thus, exhaustive testing is feasible for circuits with a limited number of inputs (not more than 25).

A modified form of exhaustive testing is *pseudo-exhaustive testing* [6.2]. It retains the advantages of exhaustive testing while significantly reducing the number of test patterns to be applied. The basic idea is to partition the circuit under test into several subcircuits such that each subcircuit has few enough inputs for exhaustive testing to be feasible for it. This concept is used in *autonomous design verification technique* proposed in Ref. 6.3. To illustrate the use of the technique, let us consider the circuit shown in Fig. 6.2(a). The circuit is partitioned into two subcircuits C_1 and C_2 as shown in Fig. 6.2(b). The functions of the two added control inputs *MC1* and *MC2* are as follows:

When $MC1 = 0$ and $MC2 = 1$, subcircuit C_2 is disabled; subcircuit C_1, shown in Fig. 6.2(c) remains; this can be tested by applying all possible input combinations at a, b, c, and d.

Similarly, when $MC1 = 1$ and $MC2 = 0$, subcircuit C_1 is disabled. In this mode of testing, the circuit functions as shown in Fig. 6.2(d) and can be tested by applying all input combinations at c, d, e, and f.

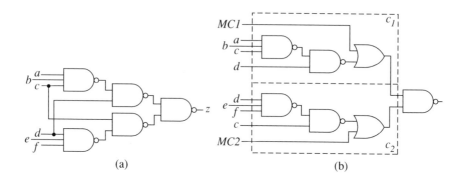

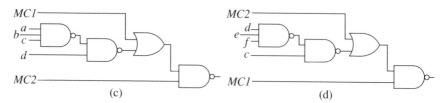

(a)   (b)

(c)   (d)

Figure 6.2 Partitioning of a circuit for autonomous testing

When $MC1 = MC2 = 0$, the circuit functions as the unmodified circuit except for the added gate delay. The advantage of the design method is that any fault in the circuit itself and the testing circuit is detectable.

A modified version of the autonomous testing, called the *verification testing*, has been proposed in Ref. 6.4. This method is applicable to multioutput combinational circuits, provided each output depends only on a proper subset of the inputs. The verification test set for a circuit is derived from its *dependence matrix*. The dependent matrix of a circuit has m rows and n columns; each row represents one of the outputs, and each column one of the inputs. An entry $[(i, j); i = 1, m, j = 1, n]$ in the matrix is 1 if the output depends on input j; otherwise, the entry is 0. To illustrate, the dependence matrix for the circuit of Fig. 6.3(a) is shown in Fig. 6.3(b). The dependence matrix is derived by tracing paths from outputs to inputs. A *partition-dependent matrix* is then formed by partitioning the columns of dependence matrix into a minimum number of sets, with each row of a set having at most one 1-entry; there may be a number of partition-dependent matrices corresponding to a dependent matrix. A partition-dependent matrix corresponding to the dependent matrix of Fig. 6.3(b) is shown in Fig. 6.3(c).

A verification test set is obtained by assigning same values to all inputs belonging to the same partition of a partition-dependent matrix; any two inputs

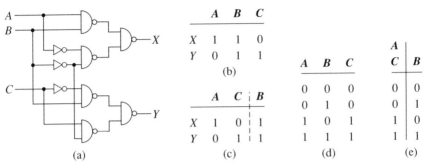

Figure 6.3 (a) Circuit under test; (b) dependence matrix; (c) partition dependent matrix; (d) verification test set; (e) reduced verification test set

belonging to different partitions receive distinct values. Figure 6.3(d) shows the verification test set for Fig. 6.3(a). A reduced verification test set can be derived from a verification test set for Fig. 6.3(a). A reduced verification test set can be derived from a verification test by removing all repetitions of identical columns. The reduced verification test set for the circuit of Fig. 6.3(a) is shown in Fig. 6.3(e).

The verification testing is an useful technique for combinational circuits or circuits that can be transformed into combinational forms during testing (e.g., LSSD structure). However, the generation of the dependence matrix, which is the most important part of this test strategy, could be a nontrivial task for circuits of VLSI complexity.

6.1.2 PSEUDO-EXHAUSTIVE PATTERN GENERATION

A combinational circuit with n inputs can be pseudoexhaustively tested with 2^w or fewer binary patterns if none of the outputs of the circuit is a function of more than w out of n inputs. For example, the following *six* test patterns can be used to pseudoexhaustively test a six-input circuit provided no output depends on more than two input variables:

$$
\begin{array}{ccccc}
1 & 1 & 1 & 1 & 1 \\
1 & 0 & 0 & 0 & 0 \\
0 & 1 & 0 & 0 & 0 \\
0 & 0 & 1 & 0 & 0 \\
0 & 0 & 0 & 1 & 0 \\
0 & 0 & 0 & 0 & 1 \\
\end{array}
$$

These test patterns can be generated by a nonprimitive polynomial, for example, $x^6 + x^4 + x^3 + x^2 + x + 1$. However, if an output of a multioutput circuit depends on more than two input variables, the derivation of minimal test patterns using a nonprimitive polynomial may not be feasible.

In general, the pseudo-exhaustive patterns needed to test an n-input and m-output combinational circuit are derived by using one of the following methods.

Syndrome driver counter

Constant weight counter

Linear feedback shift register/shift register (LFSR/SR)

Linear feedback shift register/EX–OR gates (LFSR/EX–OR)

The *syndrome driver counter* method checks if p ($< n$) inputs of the circuit under test can share the same test values with the remaining $(n - p)$ inputs [6.5]. Then the circuit can be exhaustively with 2^p inputs. To illustrate, let us consider the circuit shown in Fig. 6.4(a). No output is a function of both c and d; thus, they can share the same input signal. Also, no output is a function of both a and b; thus, they can share the same input signal. Therefore, the circuit can be tested with 2^2 ($p = 2$) combinations, as shown in Fig. 6.4(b). The syndrome driver can be a binary counter. The drawback of the method is that if the value of p is close to n, it requires as many test patterns as in exhaustive testing.

The *constant weight counter* method uses an x-out-of-Y code for exhaustively testing an m-input circuit, where x is the maximum number of input variables on which any output of the circuit under test depends [6.6]. The value of Y is chosen such that all 2^x input combinations are available from any m column of the codewords. For example, the circuit of Fig. 6.5 has six inputs and three outputs, but none of the outputs depends on more than three input variables; thus, $m = 6$ and $x = 3$.

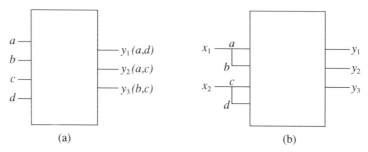

(a)   (b)

Figure 6.4   (a) Circuit under test; (b) inputs a/b and c/d tied together

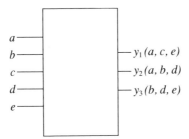

Figure 6.5   Circuit under test

If Y is chosen to be 6, the constant weight counter will generate the following 20 codewords:

000111	100011
001011	100110
001101	100101
001110	101100
010011	101010
010101	101001
010110	110001
011100	110010
011010	110100
011001	111000

The selection of *any five* columns from the preceding codewords (20 rows) will guarantee the exhaustive testing of the circuit associated with each output. In general, the constant weight test set is of minimal length. However, the complexity of the constant weight counter rapidly increases for higher x-out-of-Y code.

An alternative approach for generating a pseudoexhaustive test set is to use a combination of an LFSR (Linear Feedback Shift Register) and an SR (Shift Register) [6.7]. In an LFSR, the outputs of a selected number of stages are fed back to the input of the LFSR through an EX–OR network. An n-bit LFSR can be represented by an *irreducible* and *primitive* polynomial. If the polynomial is of degree n, then the LFSR will generate all possible $2^n - 1$ nonzero binary patterns in sequence; this sequence is termed the *maximal length sequence* of the LFSR.

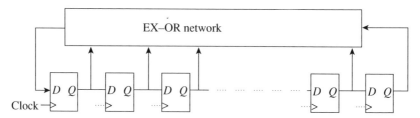

Figure 6.6   General representation of an LFSR

Figure 6.6 shows the general representation of an LFSR based on the primitive polynomial

$$P(x) = x^n + p_{n-1}x^{n-1} + \cdots + p_2x^2 + p_1x + p_0. \qquad (6.1)$$

The feedback connections needed to implement an LFSR can be derived directly from the chosen primitive polynomial. To illustrate, let us consider the following polynomial of degree 3:

$$x^4 + x + 1.$$

This can be rewritten in the form of expression (6.1):

$$P(x) = 1 \cdot x^4 + 0 \cdot x^2 + 1 \cdot x + 1 \cdot x^0.$$

Figures 6.7(a) and (b) show the four-stage LFSR constructed by using this polynomial and the corresponding maximal length sequence, respectively.

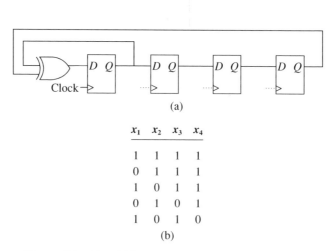

(a)

x_1	x_2	x_3	x_4
1	1	1	1
0	1	1	1
1	0	1	1
0	1	0	1
1	0	1	0

(b)

Figure 6.7   (a) A 4-bit LFSR; (b) maximal length sequence

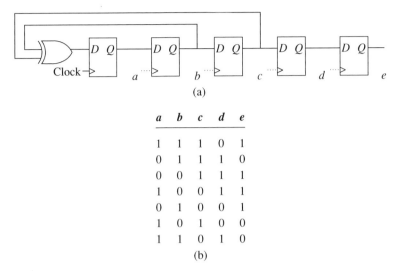

Figure 6.8 (a) LFSR/SR; (b) test patterns

Figures 6.8(a) and (b) show the combination of a 3-bit LFSR and a 2-bit SR and the resulting output sequence. Two separate starting patterns are needed, one for the LFSR and the other for the SR.

The LFSR/SR combination of Fig. 6.8 can be used to test any five-input circuit in which no output is a function of more than two input variables. This approach guarantees near minimal test patterns when the number of input variables on which any output of the C.U.T. (circuit under test) depends is less than half the total number of input variables.

A variation of the LFSR/SR approach uses a network of EX–OR gates instead of a shift register [6.8]. For example, the circuit of Fig. 6.8(a) can be modified such that d is a linear sum of a and c, and e is a linear sum of b and d. The resulting circuit and the patterns generated by it are shown in Figs. 6.9(a) and (b), respectively. In general, the LFSR/EX–OR approach produces test patterns that are very close to the LFSR/SR approach.

An alternative technique based on LFSR/SR approach called *convolved LFSR/SRs* uses an n-stage shift register and an LFSR of degree w to generate pseudo-exhaustive test patterns for a circuit with n inputs and m outputs, with no output being a function of more than w input variables [6.9].

The first step in test pattern generation using a convolved LFSR/SR is to assign *residues* R_0 through R_i to inputs 0 through i of the circuit under test. The residue of stage i is $x^i \bmod P(x)$, where $P(x)$ is the primitive polynomial used to implement the LFSR. To illustrate the computation of residues, let us consider a circuit with

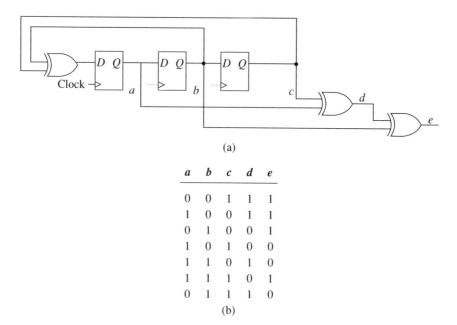

(a)

a	b	c	d	e
0	0	1	1	1
1	0	0	1	1
0	1	0	0	1
1	0	1	0	0
1	1	0	1	0
1	1	1	0	1
0	1	1	1	0

(b)

Figure 6.9   (a) LFSR/EX–OR; (b) test patterns

five inputs and five outputs as shown in Fig. 6.10. The circuit has five inputs; hence, a five-stage convolved LFSR/SR is needed. Because $w = 3$, the first three stages of the shift register are used to implement a primitive polynomial of degree 3 for example, $x^3 + x + 1$. Figure 6.11 shows the resulting convolved LFSR/SR. The residue of each stage is computed as x^i ($i = 0, \ldots, 4$) mod ($x^3 + x + 1$). For example, the residue of stage 3 is x^3 mod ($x^3 + x + 1$), that is, $x + 1$.

The next step in the test generation process is to assign residues to the inputs of the circuit under test so that for an output cone no assigned value is linearly

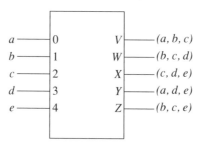

Figure 6.10   A circuit with $n = 5$, $m = 5$, and $w = 2$

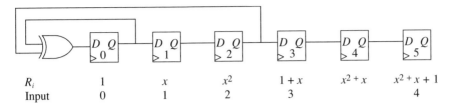

R_i	1	x	x^2	$1 + x$	$x^2 + x$	$x^2 + x + 1$
Input	0	1	2	3		4

Figure 6.11   Residue assignment for convolved LFSR/SR

dependent on already assigned values. For the circuit of Fig. 6.10, inputs a, b, c, and d can be assigned residues 1, x, x^2, and $1 + x$, respectively. However, input e cannot be assigned residue $x^2 + x$, because this will result in a residue set for output z, namely, (R_1, R_2, R_4), that is linearly dependent. This can be avoided by assigning residue R_5 to input e.

If residue R_{i+j} is selected for assignment to input $i + 1$ because residues R_{i+1}, $R_{i+2}, \ldots, R_{i+j-1}$ cannot be assigned to this input because of linear dependence, then stage $i + 1$ can be made to generate residue R_{i+j} by finding the linear sum of one or more previously assigned residues to stage $i + 1$. For the circuit under consideration, stage 4 of the convolved LFSR/SR generates the desired residue $x^2 + x + 1$ by feeding the linear sum of residues R_2 and R_3 to stage 4 as shown in Fig. 6.12.

Assuming the initial seed for the LFSR to be 110 and that for the SR to be 011, the following pseudo-random patterns are generated by the convolved LFSR/SR:

$$
\begin{array}{cccccc}
1 & 1 & 0 & 0 & 1 & 1 \\
1 & 1 & 1 & 0 & 0 & 1 \\
0 & 1 & 1 & 1 & 1 & 0 \\
1 & 0 & 1 & 1 & 0 & 1 \\
0 & 1 & 0 & 1 & 0 & 0 \\
0 & 0 & 1 & 0 & 1 & 0 \\
1 & 0 & 0 & 1 & 1 & 1 \\
\end{array}
$$

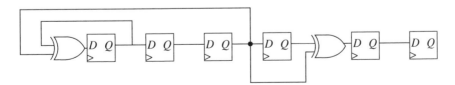

Figure 6.12   Convolved LFSR/SR

Any five columns of these patterns constitute the minimal pseudo-exhaustive test set for the output cones of the circuit of Fig. 6.10.

6.1.3 *PSEUDO-RANDOM PATTERN GENERATOR*

Pseudo-random patterns are sufficiently random in nature to replace truly random sequences. LFSRs are widely used for generating test patterns for combinational circuits because an LFSR is easy to implement. However, three related issues need to be considered in order to measure the effectiveness of pseudorandom testing [6.10].

1. Determination of the number of test patterns
2. Evaluation of the fault coverage
3. Detection of *random pattern resistant* faults

The fault coverage can be evaluated by using exhaustive fault simulation. However, pseudo-random patterns needed to test a circuit are typically large; thus, fault simulation can be expensive. A relationship between a pseudo-random test sequence of length L, and the *expected fault coverage* $E(c)$ is given in [6.11]:

$$E(c) = 1 - \sum_{k=1}^{2^n-1} \left(1 - \frac{L}{2^n}\right)^k \frac{h_k}{M},$$

where n is the number of circuit inputs, h_k $(k = 1, 2, 3, \ldots, 2^n)$ is the number of faults in the circuit that can be detected by k input vectors, and M is the total number faults in the circuit under test. It will be clear from the preceding expression that the h_k need to be known a priori to evaluate $E(c)$. The h_k for the two-input circuit of Fig. 6.13 are $(h_1, h_2, h_3, h_4) = (7, 0, 1, 0)$, as shown in Table 6.1. For complex circuits, the h_k are extremely difficult to derive, and they can only be approximated using probabilistic analysis.

A major problem associated with pseudo-random testing is that the number of patterns needed to detect a random pattern resistant fault may be very large. For example, let us consider the stuck-at-1 fault α at the input of a 10-input AND gate shown in Fig. 6.14. It is clear only test pattern $abcdefghij = 1111110111$ can detect the fault. The probability of an LFSR generating this particular pattern is 2^{-10}! Thus, a huge number of pseudo-random patterns need to be applied to a C.U.T. that may contain random pattern resistant faults, to guarantee a high fault

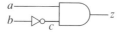

Figure 6.13 A two-input circuit

Table 6.1   **Fault Detectability for the Circuit of Fig. 6.13**

Test	Fault							
ab	*a* s-a-0	*a* s-a-1	*b* s-a-0	*b* s-a-1	*c* s-a-0	*c* s-a-1	*z* s-a-0	*z* s-a-1
00		×						×
01								×
10	×			×	×		×	
11			×			×		×
k	1	1	1	1	1	1	1	3

$$h_1 = 7, \quad h_2 = 0, \quad h_3 = 1, \quad h_4 = 0$$

coverage. The inherent weakness of pseudo-random patterns as far as the detection of random pattern resistant faults is concerned arises because each bit in such a pattern has a probability of 0.5 of being either 0 or 1. If, instead of generating patterns with uniform distribution of 0s and 1s, a biased distribution is used, there is a higher likelihood of finding test patterns for random pattern resistant faults. This is the principle of the *weighted test generation* technique proposed in Ref. 6.12.

An alternative way of generating pseudorandom patterns is to use a *cellular automation (CA)*. A CA consists of a number of identical cells interconnected spatially in a regular manner [6.13]. Each cell consists of a *D* flip-flop and combinational logic that generates the next state of the cell. Several rules may be used to compute the next state of a cell. For example, rules 90 and 160 are defined as

$$\text{Rule 90:} \quad y_i(t + 1) = y_{i-1}(t) \oplus y_{i+1}(t),$$

$$\text{Rule 160:} \quad y_i(t + 1) = y_{i-1}(t) \oplus y_i(t) \oplus y_{i+1}(t),$$

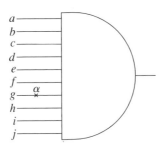

Figure 6.14   A 10-input AND gate with a stuck-at-1 fault

where $y_i(t)$ denotes the state of cell i at time t. It has been shown in Ref. 6.14 that, by combining these two rules, it is possible to generate a sequence of maximal length $2^s - 1$, where s is the number of cells in a CA. To illustrate, let us consider the four-cell CA shown in Fig. 6.15(a). Assuming the initial state of the CA to be $abcd = 0101$, the maximal length sequence for the CA is shown in Fig. 6.15(b). The implementation of the CA is shown in Fig. 6.15(c). Note that

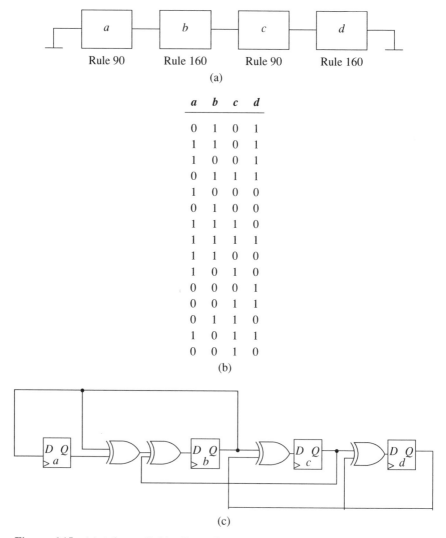

a	b	c	d
0	1	0	1
1	1	0	1
1	0	0	1
0	1	1	1
1	0	0	0
0	1	0	0
1	1	1	0
1	1	1	1
1	1	0	0
1	0	1	0
0	0	0	1
0	0	1	1
0	1	1	0
1	0	1	1
0	0	1	0

(b)

(c)

Figure 6.15   (a) A four-cell CA; (b) maximal length sequence; (c) implementation of the four-cell CA

a 4-bit LFSR implementing a primitive polynomial of degree 3 will also generate a sequence of length 16. CAs based on rules 90 and 160 can generate all primitive and irreducible polynomials of a given degree [6.15]. Also, CAs do not require long feedback, which results in smaller delays and efficient layouts.

6.1.4 DETERMINISTIC TESTING

Traditional test generation techniques may also be used to generate test patterns that can be applied to the circuit under test when it is in BIST mode. The test patterns and the corresponding output responses are normally stored in a ROM. If the output responses of the circuit under test do not match the expected responses when the stored test patterns are applied, the presence of a fault(s) is assumed. Although in principle this is a satisfactory approach for fault detection, it is rarely used because of the high overhead associated with storing test patterns and their responses.

6.2   Output Response Analysis

As stated earlier, BIST techniques usually combine a built-in binary pattern generator with circuitry for compressing the corresponding response data produced by a C.U.T. The compressed form of the response data is compared with a known fault-free response. Several compression techniques that can be used in a BIST environment have been proposed over the years; these include

1. Transition count
2. Syndrome checking
3. Signature analysis

6.2.1 TRANSITION COUNT

The *transition count* is defined as the total number of transitions from $1 \to 0$ and $0 \to 1$ in a response sequence for a given input sequence. For example, if a response sequence $Z = 10011010$, then the transition count $c(Z) = 5$. Thus, instead of recording the entire output response sequence, only the transition count is recorded. The transition count is then compared with the expected one, and if they differ, the C.U.T. is declared faulty [6.16].

Figure 6.16(a) shows the response sequences and the corresponding transition counts at various nodes of a circuit resulting from the application of a test sequence of length 5. Let us suppose there is a fault α s-a-0 in the circuit (Fig. 6.16(b)). The presence of the fault changes the transition counts at certain

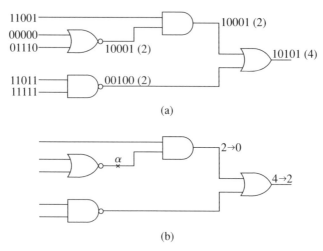

(a)

(b)

Figure 6.16   (a) Response to test sequence of length 5; (b) changes in transition counts

nodes in the circuit (shown by arrows). As can be seen in the diagram, the transition count at the output node changes from 4 to 2, resulting in the detection of the fault α s-a-0.

The main advantage of transition counting is that it is not necessary to store either the correct response sequence or the actual response sequence at any test point; only the transition counts are needed. Clearly, this results in the reduction of data storage requirements. However, this data compression may give rise to fault-masking errors. This is because most transition counts correspond to more than one sequence; for example, the transition count 2 is generated by each of the following 6-bit sequences: 01110, 01100, 01000, 00110, 01000, and 00010. Hence, there is a possibility that a fault sequence will produce the same transition count as the good sequence and, therefore, go undetected. However, as the sequence length increases, the hazard of fault masking diminishes.

6.2.2 SYNDROME CHECKING

The syndrome of a Boolean function is defined as $S = K/2^n$, where K is the number of minterms realized by the function and n is the number of input lines [6.17]. For example, the syndrome of a three-input AND gate is $\frac{1}{8}$ and that of a two-input OR gate is $\frac{3}{4}$. Because the syndrome is a functional property, various realizations of the same function have the same syndrome.

The input–output syndrome relation of a circuit having various interconnected blocks depends on whether the inputs to the blocks are disjoint or conjoint, as well as on the gate in which the blocks terminate. For a circuit having two blocks with unshared inputs, if S_1 and S_2 denote the syndromes of the functions realized by the blocks 1 and 2, respectively, the input–output syndrome relation S for the circuit is

Terminating Gate	Syndrome Relation S
OR	$S_1 + S_2 - S_1 S_2$
AND	$S_1 S_2$
EX–OR	$S_1 + S_2 - 2 S_1 S_2$
NAND	$1 - S_1 S_2$
NOR	$1 - (S_1 + S_2 - S_1 S_2)$

If blocks 1 and 2 have shared inputs and realize the function F and G, respectively, then the following relations hold:

$$S(F + G) = S(F) + S(G) - S(FG),$$

$$S(FG) = S(F) + S(G) - S(\overline{F}\,\overline{G}) - 1,$$

$$S(F \oplus G) = S(\overline{F}G) + S(F\overline{G}).$$

As an example, let us find the syndrome and the number of minterms realized by the fan-out-free circuit of Fig. 6.17. We have $S_1 = \frac{3}{4}$ and $S_2 = \frac{1}{4}$. Hence, $S_3 = 1 - S_1 S_2 = \frac{13}{16}$, and $K = S \cdot 2^n = 13$. Table 6.2 lists the syndrome of the fault-free circuit of Fig. 6.18, and the syndromes in the presence of fault α s-a-0 and the fault β s-a-1.

6.2.3 SIGNATURE ANALYSIS

Signature analysis technique is pioneered by Hewlett-Packard Ltd. that detects errors in data streams caused by hardware faults [6.18]. It uses a data compaction technique to reduce long data streams into a unique code called the *signature*. Signatures can be created from the data streams by feeding the data into an n-bit

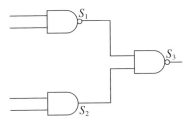

Figure 6.17 A fan-out-free circuit

Table 6.2 **Fault-Free and Faulty Syndromes**

$x_1x_2x_3$	Output response		
	Fault-free	α s-a-1	β s-a-0
0 0 0	1	1	1
0 0 1	1	1	1
0 1 0	1	0	0
0 1 1	0	0	0
1 0 0	1	0	0
1 0 1	0	0	0
1 1 0	1	0	0
1 1 1	0	0	0
Syndrome	5/8	2/8	2/8

LFSR. The feedback mechanism consists of EX–ORing selected taps of the shift register with the input serial data as shown in Fig. 6.19. After the data stream has been clocked through, a residue of the serial data is left in the shift register. This residue is unique to the data stream and represents its signature. Another data stream may differ by only one bit from the previous data stream, and yet its signature is radically different from the previous one. To form the signature of a data stream, the shift register is first initialized to a known state and then shifted using the data stream; normally, the all-0s state is chosen as the initial state.

Figure 6.20(a) shows a simplified 4-bit signature generator. If a 1 is applied to the circuit, the EX–OR gate will have output 1. The next clock pulse will shift the gate output into the first stage of the register and 0s from the preceding stages into the second, third, and fourth stages, which leaves the register containing 1000, that is, in state 8. From the state diagram of Fig. 6.20(b), the register contents or signatures can be identified for any data stream.

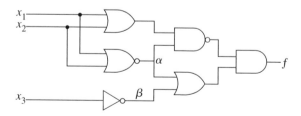

Figure 6.18 Circuit under test

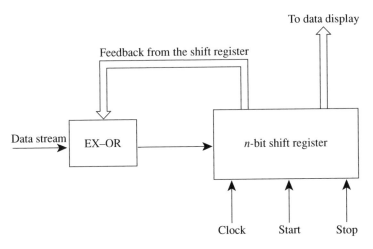

Figure 6.19   Signature analyzer circuit

An n-stage signature generator can generate 2^n signatures. However, many input sequences can map into one signature. In general, if the length of an input sequence is m and the signature generator has n stages, then 2^m input sequences map into 2^n signatures. In other words, 2^{m-n} input sequences map into each signature. Only one out of 2^m possible input sequences is error-free and produces the correct signature. However, any one of the remaining $2^{m-n} - 1$ sequences may also map into the correct signature. This mapping gives rise to *aliasing*, that is, the signature generated from the faulty output response of a circuit may be identical to the signature obtained from the fault-free response. In other words, the presence of a fault in the circuit is masked. The probability P that an input sequence has deteriorated into another having the same signature itself is

$$P = \frac{2^{m-n} - 1}{2^m - 1}. \tag{6.2}$$

P is calculated on the assumption that any of the possible input sequences of a given length may be good or faulty. Expression (6.2) reduces to

$$P = \frac{1}{2^n} \qquad \text{for } m \gg n.$$

Thus, the probability of aliasing will be low if a signature generator has many stages and hence is capable of generating a large number of signatures. For example, the 16-bit signature generator shown in Fig. 6.21 can generate 66,636

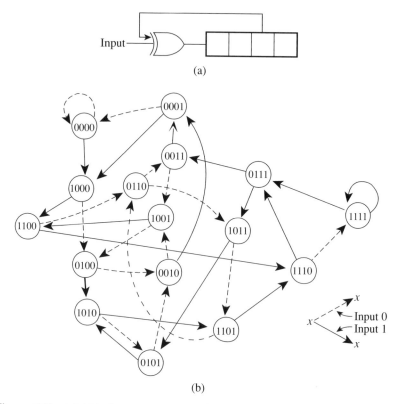

(a)

(b)

Figure 6.20 (a) 4-bit signature generator; (b) state diagram of the signature generator

signatures, and hence the probability that two input sequences will produce the same signature is 0.002%.

The error detection properties of the signature analysis technique are as follows:

1. The probability that two identical input sequences will produce the same signature is 1.

2. The probability that input sequences will produce the same signature if they differ precisely by one bit is 0. For example, let us consider two long input sequences, one differing from the other by only one bit. As the error bit gets shifted into the 16-bit register of Fig. 6.21, it has four chances to change the signature register's input before it overflows the register (after 16 clock cycles) and disappears. The effect of the erroneous bit continues to propagate around the feedback network, changing the signature. For a single bit error, therefore, no other error bit comes along

Figure 6.21 A 16-bit signature generator

to cancel the feedback's effect, and so signature analysis is bound to catch the single bit error. Single bit errors, incidentally, are typical of transient errors that occur in VLSI devices.

A signature corresponding to the output sequence produced by a C.U.T. is usually created, as discussed previously, by feeding the sequence serially into the feedback line of an LFSR via an additional EX–OR gate. A signature can also be obtained by feeding a subset of the output sequence in parallel when a *multiple-input signature register (MISR)* is used. A k-bit MISR can compact an m ($\gg k$)-bit output sequence in m/k cycles. Thus, a MISR can be considered as a *parallel signature analyzer*. Figure 6.22 shows an 8-bit MISR.

6.3 Circular BIST

The circular BIST technique is a special form of BIST technique based on pseudo-random testing [6.19]. Figure 6.23(a) shows a simplified configuration of circular BIST. The block denoted as *Logic Module (LM)* is the circuit under test without certain selected memory elements. These memory elements are replaced by special flip-flop cells, called *circular flip-flops (CFFs)*. A circular flip-flop uses a 2-to-1 multiplexer and an EX–OR gate as shown in Fig. 6.23(b). It can operate in two different modes, functional and BIST, as indicated in Table 6.3. Two special additional control lines can be incorporated into a CFF as shown in

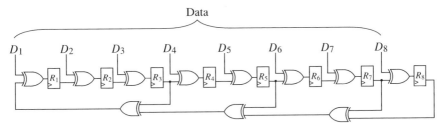

Figure 6.22 Multiple-input signature register (MISR)

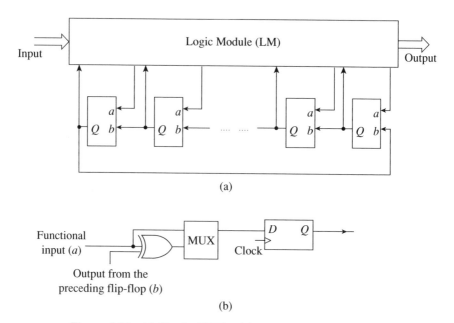

(a)

(b)

Figure 6.23   (a) Circular BIST architecture; (b) circular flip-flop

Fig. 6.24 to provide four modes of operation [6.20]. Table 6.4 identifies these modes of operation.

As it will be clear from Fig. 6.23, the circular BIST technique configures certain flip-flops into a MISR, where the LM constitutes the feedback logic. The test process begins with the initialization of the LM block and the resetting of all CFFs by using the CFF reset mode. Alternatively, a separate partial scan path to initialize the CFFs can be made by selecting the *shift* mode. The contents of the CFFs constitute a test pattern for the LM block. Once the CFFs are clocked, each LM output is EX–ORed with the output of the preceding flip-flop. The resulting output pattern is considered as the signature corresponding to the first input pattern, and it is used as the next pseudo-random test pattern for the circuit under

Table 6.3   **CFF Function Table**

Mode	Output
Functional	a
BIST	$a \oplus b$

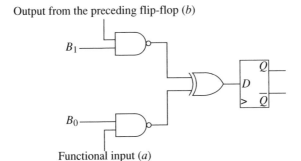

Output from the preceding flip-flop (b)

B_1

B_0

Functional input (a)

Figure 6.24 Modified CFF

test. This process of applying a test pattern and generating signature data is continued for a specified number of clock cycles. The final signature is used for the verification of the faulty/fault-free status of the circuit under test. An additional signature generator may also be included in the chain of CFFs.

The advantage of the circular BIST technique is that it provides high fault coverage and low hardware overhead. However, the set of test patterns generated depends on the function of the circuit under test. If certain test patterns are not generated, a number of faults may not be detected, thereby lowering the fault coverage compared with conventional pseudo-random testing. It is possible, however, to improve the fault coverage by configuring the CFFs in a partial scan path and applying deterministic test patterns via this path [6.21]. Another problem associated with the circular BIST is the selection of flip-flops to be used as CFFs. Obviously, the fewer the CFFs, the lower the hardware overhead will be. This, however, has a negative impact on the circuit testability because a small number of CFFs may not allow generation of enough random patterns to satisfy the fault coverage requirement. Several heuristic criteria have been proposed in Ref. 6.21 for selecting flip-flops to be used as CFFs.

Table 6.4 **Modified CFF Function Table**

B_0	B_1	Mode	Output
0	0	Reset	0
0	1	Shift	b
1	0	Functional	a
1	1	BIST	$a \oplus b$

6.4 BIST Architectures

Over the years, several BIST architectures have been proposed by researchers in industry and universities. We discuss some of these in this section.

6.4.1 BILBO (Built-In Logic Block Observer)

In the BILBO structure, the scan-path technique is combined with signature analysis [6.22]. It uses a multipurpose module, called a BILBO, that can be configured to function as an input test pattern generator or an output signature analyzer. This is composed of a row of flip-flops and some additional gates for shift and feedback operations. Figure 6.25 shows the logic diagram of a BILBO. The two control inputs B_1 and B_2, are used to select one of the four function modes:

Mode 1 $B_1 = 0, B_2 = 1$. All flip-flops are reset.

Mode 2 $B_1 = 1, B_2 = 1$. The BILBO behaves as a latch. The input data $x_1, \ldots, x_n$ can be simultaneously clocked into the flip-flops and can be read from the Q and $\overline{Q}$ outputs.

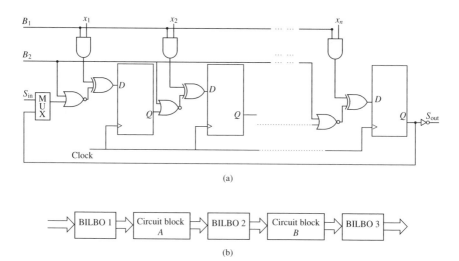

(a)

(b)

Figure 6.25 (a) Logic diagram of a BILBO; (b) BILBO-based BIST architecture

Mode 3 $B_1 = 0, B_2 = 0$. The BILBO acts as a serial shift register. Data are serially clocked into the register through S_{in} while the register contents can be simultaneously read at the parallel Q and $\overline{Q}$ outputs, or clocked out through the serial output S_{out}.

Mode 4 $B_1 = 1, B_2 = 0$. The BILBO is converted into a multiple-input signature register. In this mode, it may be used for performing parallel signature analysis, or for generating pseudo-random sequences. The latter application is achieved by keeping $x_1, \ldots, x_n$ at fixed values.

Figure 6.25(b) shows the BILBO-based BIST architecture for two cascaded circuit blocks A and B. BILBO 1 in this structure is configured as a pseudo-random pattern generator, the outputs of which are applied as test inputs to circuit block A. BILBO 2 is configured as a parallel signature register and receives its inputs from circuit block A. Similarly, BILBOs 2 and 3 should be configured to act as a pseudo-random pattern generator and a signature register, respectively, to test circuit block B. It should be clear that circuit blocks A and B cannot be tested in parallel, because BILBO 2 has to be modified to change its role during the testing of these blocks.

A modified version of the conventional BILBO structure is shown in Fig. 6.26(a) [6.23]. In addition to normal, serial scan, and MISR function, the modified BILBO can also function as an LFSR, thus generating pseudo-random patterns (Fig. 6.26(b)). The modified BILBO can be used for simultaneous testing of pipeline structure. For example, in Fig. 6.26(c) circuit blocks A and C can be simultaneously tested by operating BILBOs 1 and 3 in the LFSR mode and BILBOs 2 and 4 in the MISR mode. Circuit block B can be tested individually by making BILBOs 2 and 3 operate in the LFSR and MISR modes, respectively.

6.4.2 STUMPS (Self-Testing Using an MISR and Parallel Shift Register Sequence Generator)

STUMPS uses multiple serial scan paths that are fed by a pseudo-random number generator as shown in Fig. 6.27 [6.24]. Each scan path corresponds to a segment of the circuit under test and is fed by a pseudo-random number generator. Because the scan paths may not be of same length, the pseudo-random generator is run till the largest scan path is loaded. Once data has been loaded into the scan paths, the system clock is activated. The test results are loaded into the scan paths and then shifted into the MISR.

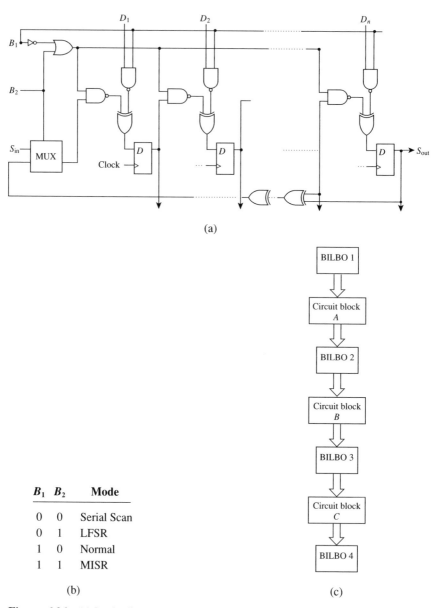

(a)

B_1	B_2	Mode
0	0	Serial Scan
0	1	LFSR
1	0	Normal
1	1	MISR

(b)

(c)

Figure 6.26 (a) Logic diagram of a modified BILBO; (b) operating mode; (c) simultaneous testing of pipeline structure

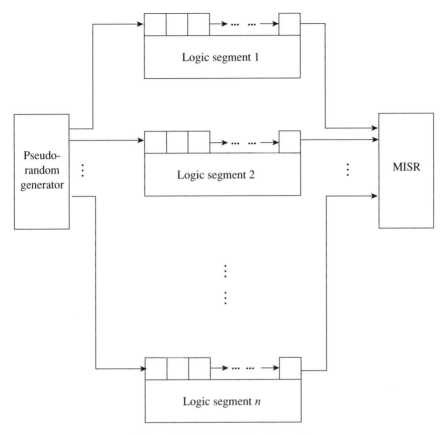

Figure 6.27   STUMPS configuration

6.4.3   LOCST (LSSD On-Chip Self-Test)

LOCST combines pseudo-random testing with LSSD-based circuit structure
[6.25]. The inputs are applied via boundary scan cells; also, the outputs are obtained via boundary scan cells. The input and the output boundary scan cells
together with the memory elements in the circuit under test form a scan path.
Some of the memory elements at the beginning of the scan path are configured
into an LFSR for generating pseudo-random numbers. Also, some memory elements at the end of the scan path are configured into another LFSR, which functions as a signature generator. Figure 6.28 shows the LOCST configuration.

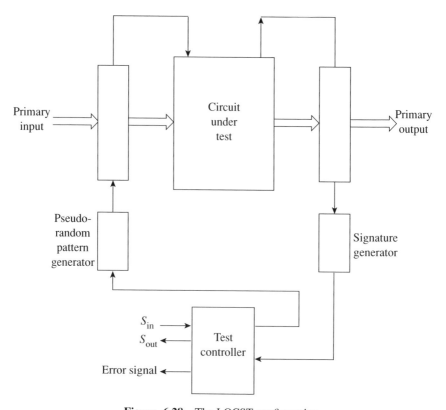

Figure 6.28   The LOCST configuration

The test process starts by serially loading the scan path consisting of input boundary scan cells, with pseudo-random patterns generated by the LFSR. These patterns are applied to the combinational part of the C.U.T., and the resulting output patterns are loaded in parallel into the scan path consisting of output boundary scan cells. These output bits are then shifted into the signature register. The resulting signature is the compared with the reference signature for verification purposes.

References

6.1   Williams, T. W., ''Design for testability,'' in *Computer Design Aids for VLSI Circuits* (Eds. P. Antogneti, D. O. Pederson, and H. de Mann), Martinus Nijhoff, 359–416 (1986).

6.2   McCluskey, E. J., *Logic Design Principles: With Emphasis on Testable Semicustom Circuits*, Prentice Hall, (1986).

6.3   McCluskey, E. J., and S. B. Nesbet, "Design for autonomous test," *IEEE Trans. Circuits Syst.*, 1070–1079 (November 1981).

6.4   McCluskey, E. J., "Verification testing—a pseudoexhaustive test technique," *IEEE Trans. Comput.*, 541–546 (June 1984).

6.5   Barzilai, Z., J. Savir, G. Markowsky, and M. G. Smith, "The weighted syndrome sums approach to VLSI testing," *IEEE Trans. Comput.*, 996–1000 (December 1981).

6.6   Tang, D. L., and L. S. Woo, "Exhaustive pattern generation with constant weight vectors," *IEEE Trans. Comput.*, 1145–1150 (December 1983).

6.7   Barzilai, Z., D. Coopersmith, and A. Rosenberg, "Exhaustive bit pattern generation in discontiguous positions with applications to VLSI testing," *IEEE Trans. Comput.*, 190–194 (February 1983).

6.8   Akers, S. B., "On the use of linear sums in exhaustive testing," *Proc. 15th Annual Symp. Fault-Tolerant Computing*, 148–153 (June 1985).

6.9   Srinivasan, R., S. K. Gupta, and M. A. Breuer, "Novel test pattern generators for pseudo-exhaustive testing," *Proc. Intl. Test Conf.*, 1041–1050 (1993).

6.10 Agrawal, V. D., C. R. Kime, and K. K. Saluja, "A tutorial on built-in-self-testing, Part 1: Principles," *IEEE Design and Test of Computers*, 73–82 (March 1993).

6.11 Wagner, K. D., C. K. Chin, and E. J. McCluskey, "Pseudorandom testing," *IEEE Trans. Comput.*, 332–343 (March 1987).

6.12 Wunderlich, H. J., "Self test using equiprobable random patterns," *Proc. Intl. Symp. Fault-Tolerant Computing*, 258–263 (1987).

6.13 Wolfram, S., "Statistical mechanics of cellular automata," *Rev. Mod. Phys.* (July 1985).

6.14 Hortensius, P. D., R. D. McLeod, and B. W. Podaima, "Cellular automata circuits for built-in-self-test," *IBM Jour. Res. and Dev.*, 389–405 (March–May 1990).

6.15 Serra, M., T. Slater, J. C. Muzio, and D. M. Miller, "The analysis of one-dimensional cellular automata and their aliasing properties," *IEEE Trans. on CAD*, 767–778 (July 1990).

6.16 Hayes, J. P., "Transition count testing of combinational logic networks," *IEEE Trans. Comput.*, 613–620 (June 1976).

6.17 Savir, J., "Syndrome-testable design of combinational circuits," *IEEE Trans. Comput.*, 442–451 (June 1980).

6.18 Hewlett-Packard Corp., *A Designer's Guide to Signature Analysis*, Application note 222 (April 1977).

6.19 Krasniewski, A., and S. Pilarski, "Circular self test path: A low cost BIST technique," *Proc. ACM/IEEE Design Automation Conf.*, 407–415 (1987).

6.20 Stroud, C. E., "Automated BIST for sequential logic synthesis," *IEEE Design and Test of Computers*, 22–32 (1979).

6.21 Pradhan, M. M., E. J. O'Brien, S. L. Lam, and J. Beausang, "Circular BIST with partial scan," *Proc. Intl. Test Conf.*, 719–729 (1988).

6.22 Koenamann, B. J. Mucha, and G. Zwiehoff, "Built-in logic block observation techniques," *Proc. Intl. Test Conf.*, 37–41 (1979).

6.23 Agrawal, V. D., C. R. Kime, and K. K. Saluja, "A tutorial on built-in-self-testing, Part 2: Applications," *IEEE Design and Test of Computers*, 69–77 (June 1993).

6.24 Bardell, P. H., W. H. McAnney, and J. Savir, *Built-In Test for VLSI Pseudorandom Techniques*, John Wiley and Sons, (1987).

6.25 LeBlanc, J. J., "LOCST: A built-in-self-test technique," *IEEE Design and Test of Computers*, 42–52 (November 1984).

Chapter 7 | Testable Memory Design

Random Access Memories (RAMs) allow data to be written and retrieved from physical locations in any order with time to access a given location being constant. Semiconductor RAMs are commonly divided into two types: *dynamic* and *static*. A dynamic RAM (DRAM) uses a single transistor to create an extremely small capacitor for storing a data bit. Because the charge in the capacitor, that is, the memory cell, leaks away very rapidly, it must be refreshed periodically to maintain the original charge. In a static RAM (SRAM), each bit of data is stored in a cell composed of several transistors. Because a DRAM device uses fewer cells, it can have a higher density that a SRAM. However, SRAMs are in general faster than DRAMs. Both types of RAMs are volatile, that is, the stored data is lost when the power supply voltage is turned off.

The density of RAM chips has been increasing dramatically in recent years. With the increasing density, the testing of such chips has been progressively more difficult. Also, many ASIC (Application Specific Integrated Circuit) chips included memories with varying word lengths. The testing of such embedded memories is even more complex than testing memory chips because of the lack of accessibility of all signals via the ASIC pins. A practical solution is to use built-in self-testing for the embedded memories.

7.1 RAM Fault Models

A variety of faults may occur in a RAM. Before considering how faults in RAMs can be represented using functional models, let us consider a generic RAM architecture, shown in Fig. 7.1. The memory array consists of individual memory

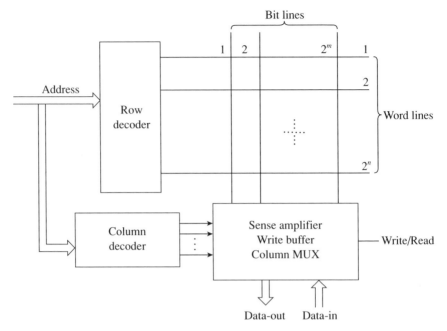

Figure 7.1   RAM organization

cells arranged in array of rows and columns. In this architecture, there are 2^n rows and 2^m columns; each row is called a *word line*, and each column a *bit line*. A memory cell can be accessed by selecting the corresponding word line and the corresponding bit line. A 1-out-of-2^n row decoder is used to select a word line, and a 1-out-of-2^m column decoder for selecting a bit line. During a read operation, the content of an accessed memory cell is transferred to the data output via a buffer called a sense amplifier. In the case of a write operation, the data input is transferred to an accessed cell via a write buffer.

Faults in the address decoder can result in the following [7.1]:

1. No memory location is accessed.

2. An undesired location is accessed.

3. More than one location are accessed simultaneously.

A defect in the memory array can manifest itself as a single fault (*stuck-at/ stuck-open fault, transition fault*), or as a fault that corrupts a memory cell that corrupts another cell [7.2]. The latter type of fault is known as a *coupling fault*.

A memory cell is said to be stuck-at-0 (or 1) if the value of the cell cannot be changed from its current logical value. A memory cell is stuck open if it cannot be accessed. If a memory cell fails to make a transition from a 0 to 1, or vice versa, it is assumed to have a transition fault. It is possible for the same cell to exhibit both types of transition faults.

A pair of memory cells p and q may be *coupled* such that a transition from 0 to 1 (1 to 0) in cell p changes the content of cell q from 1 to 0 (0 to 1); cell p is the *coupling* cell and cell q is the *coupled* cell. The coupling between two cells results in the coupling fault. Note that the content of the coupled cell is changed only at the time of transition in the coupling cell. Thus, any subsequent write operation can remove the coupling and change the content of the coupled cell. A coupling fault can be one of the following types: *inversion coupling fault, idempotent coupling fault*, or *state coupling fault*. An inversion coupling fault results if a 0-to-1 or a 1-to-0 write operation into cell p inverts the content of cell q. An idempotent coupling fault occurs if a write operation into cell p forces the content of cell q to be 0 or 1. Cell q is said to be state-coupled to cell p if the content of cell p rather than a transition in it determines the content of cell q. As long as the coupling cell p is in a particular state (0 or 1), the coupled cell q is forced to a certain value. Thus, a state coupling fault, unlike the other two types of coupling faults, is not activated by a write operation. It has been observed that local defects, also known as spot defects, in the memory cell array contribute to about 50% of stuck-at faults, 16% of stick-open faults, and 12% of state coupling faults [7.2]. Idempotent coupling faults and transition faults occur only if the sizes of spot defects are large.

A memory cell is said to exhibit *pattern sensitivity* if its content changes by a pattern of 0s and 1s, a transition from a 0 to a 1, or vice-versa, in its adjacent memory cells. This phenomenon arises due to interactions between adjacent memory cells, and therefore it is more likely to happen in high density memory chips because of the close proximity of the memory cells. Pattern sensitivity can be represented by a *static* or a *dynamic fault model*. In the static pattern-sensitive fault model, the content of a memory cell is assumed to be forced to a certain state when its adjacent cells contain a certain pattern. A dynamic path-sensitive fault model, on the other hand, assumes that the content of a memory cell changes its state due to *changes* in the contents of its adjacent cells.

The content of a memory cell is more likely to be affected by the contents of other cells in its row and column rather than by the activities in its adjacent cells. This is because the cells of the same row share a common word line, and the cells in the same column share a common bit line [7.3]. If the content of a memory cell is sensitive to the *weight*, that is, the number of 1s in cells in its row and column neighborhood, the *row/column weight-sensitive fault model* can be used to represent such pattern sensitivity.

The functional fault models of RAMs just discussed do not cover faults that are internal to chips and time dependent; such faults are called *dynamic faults* [7.4]. These include *sense amplifier recovery*, *imbalance faults*, *write recovery*, and *data retention faults*. A sense amplifier recovery occurs if the sense amplifier in a RAM gets saturated after reading a long stream of similar data bits and incorrectly reads a bit with opposite value. Imbalance faults may occur in dynamic RAMs if the sense amplifier reads the content of a memory cell erroneously because most of the other cells on the same bit line contain opposite values. A write recovery may occur if a write operation at a location followed by a read or write operation at a different location uses the first location for the second operation due to slow decoding of the address. A data retention fault in a memory cell forces it to lose its content after some units of time.

7.2   Test Algorithms for RAMs

Several test algorithms for RAMs have been proposed over the years. The complexity of these algorithm vary from $O(n)$ to $O(n^2)$, where n is the number bits in the RAM chips. We briefly discuss a few of these that have been used in industry and are being employed in modified form in RAMs with BIST features.

7.2.1   GALPAT (Galloping 0s and 1s)

In this technique, proposed by Breuer and Friedman [7.5], all the cells in the memory array are initialized to 0s. An address is then selected as the reference address, and its content is changed from 0 to 1 (assuming each location stores 1-bit data). Next, another location is accessed, and its content is read to check if it is 0. The read and verification operation is continued for all memory locations. After that, the reference location is read again to verify its content is 1, and then a 0 is written into it. The entire process is repeated by selecting each location of the memory as a reference location in turn, and changing its content from a 0 to a 1. Assuming a read and compare is a single operation, the execution time of the GALPAT is of the order N^2, where N is the number of locations in the memory under test.

7.2.2 WALKING 0s AND 1s

This technique, as does GALPAT, initializes all memory locations to 0s [7.5]. A 1 is written into a selected reference location. All other locations are then read in sequence to verify that each contains a 0. This is done to ensure the content of none of these locations is disturbed by the write operation in the reference location. Next, a 0 is written into the reference location, and a 1 into another location selected as the reference location. The entire process is then repeated. Note that, unlike in the GALPAT, a reference location is read only once in this technique. The execution time of the technique is of the order $2N^2$.

7.2.3 MARCH TEST

The March technique proposed by Nair [7.6] and later improved by Suk and Reddy [7.7] initializes the memory locations with 0s. then, each memory location is read in ascending order to check if it contains a 0 (assuming each location stores 1-bit data); a 1 is then written into it. Once the read/write process is complete, each location is accessed in descending order to check that it contains a 1, and then a 0 is written into it. Figure 7.2 shows the status of a 4 × 1 memory at the end of each step. Because the memory locations are accessed in ascending order, any direct coupling between a memory cell in the currently accessed location and a higher location is detected when the higher location is read. Similarly, accessing the memory locations in a descending order detects any coupling effect

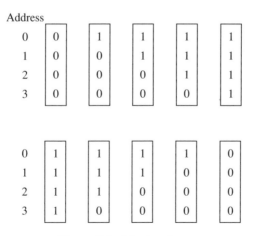

Figure 7.2 March pattern

a higher location may have on memory cells of lower locations. Also, because all cells in a location are set to both 0 and 1, a stuck-at fault in a cell is detected. The execution time of the algorithm is of order N, where N is the number of memory locations.

7.2.4 CHECKERBOARD TEST

This test algorithm selects the memory locations in a sequence, and it writes 0s and 1s into alternate memory cells in a location (a multibit word is assumed). The content of a location is read after the write operation to check whether any stored bit is disturbed. After all locations have been written into and read, the entire process is repeated by interchanging 0s and 1s. Figure 7.3 shows two patterns for the checkerboard testing of a 4×4 memory array. The test time needed by the algorithm is $O(N)$. It covers stuck-at faults as well as data retention faults. However, decoder faults are not detected.

7.3   Detection of Pattern-Sensitive Faults

As mentioned earlier, pattern-sensitive faults are more likely to occur in high density memory chips. Kinoshita and Saluja [7.8] proposed a technique for detecting static pattern-sensitive faults in memory arrays. The cells under test in a memory array, called the *base cells*; they are labeled with ∗s, and all other cells are labeled A, B, C, and D. Figure 7.4(a) shows an 8×8 memory array with labeled cells. Note that all base cells except those on the first and the last row, and on the first and the last column are surrounded by four cells (A, B, C, D). The test technique operates in two phases. During the first phase, all cells are initialized to 0s, and then all possible, that is, 16, binary patterns are sequentially written into cells labeled A, B, C, D. After each write operation, the contents of

0	1	0	1
1	0	1	0
0	1	0	1
1	0	1	0

1	0	1	0
0	1	0	1
1	0	1	0
0	1	0	1

Figure 7.3   Checkerboard pattern

*	A	*	B	*	A	*	B
C	*	D	*	C	*	D	*
*	B	*	A	*	B	*	A
D	*	C	*	D	*	C	*
*	A	*	B	*	A	*	B
C	*	D	*	C	*	D	*
*	B	*	A	*	B	*	A
D	*	C	*	D	*	C	*

(a)

A	*	B	*	A	*	B	*
*	C	*	D	*	C	*	D
B	*	A	*	B	*	A	*
*	D	*	C	*	D	*	C
A	*	B	*	A	*	B	*
*	C	*	D	*	C	*	D
B	*	A	*	B	*	A	*
*	D	*	C	*	D	*	C

(b)

Figure 7.4   Test for static pattern sensitive faults: (a) Phase 1; (b) Phase 2

base cells are checked for possible pattern sensitivity. The second phase of the test process relabels the memory array as shown in Fig. 7.4(b); then, the writing and the reading operations are performed as in the first phase. The memory array is next initialized with all 1s, and both test phases are repeated.

Franklin *et al.* [7.3] have proposed an algorithm for detecting row/column pattern-sensitive faults. This algorithm also detects static pattern-sensitive faults, stuck-at faults, and coupling faults. The first step of the algorithm is to initialize the memory array into A(1) state, that is, all the cells in the memory contain 1. Next, each cell of the first row is read, and a 0 is written into it. This process is continued for each row, at the end of which the memory is in A(0) state, that is, all the cells in the memory contain 0. Each cell of row 0, row 1, . . . , row $n - 1$ is then read in sequence, and a 1 is written into it. After all the cells in the memory array have been written into, the memory will be in A(1) state. Next, a row *weight* 1 is applied to the corner cell (0,0); the row (column) weight of a cell is the number of 1s in other cells of the same row (column). This is done by *flipping* the $(n - 1)$th column first, and then flipping every row in the memory array as described previously. A flip operation involves writing the complement of the current content of a row (or a column) into itself. Figure 7.5 illustrates how a row weight of 1 is applied to location (0,0) of a 4 × 4 memory array. It is assumed that the memory is in state A(1) (Fig. 7.5(a)). Figure 7.5(b) shows the status of the memory array after column 3 is flipped. Next, each row of the memory is flipped one at a time. Figures 7.5(c)–(f) show the memory state after the flipping of rows 0, 1, 2, and 3, respectively. Note that each cell in row 3 except cell 3 has row weight 1, and column weight varying from 0 to 3 when the content of the cell is 0 (Figs. 7.5(c)–(f)). The row weight of cell 3 in row 0 is 0 and the column

	0	1	2	3
0	1	1	1	1
1	1	1	1	1
2	1	1	1	1
3	1	1	1	1

(a)

	0	1	2	3
0	1	1	1	0
1	1	1	1	0
2	1	1	1	0
3	1	1	1	0

(b)

	0	1	2	3
0	0	0	0	1
1	1	1	1	0
2	1	1	1	0
3	1	1	1	0

(c)

	0	1	2	3
0	0	0	0	1
1	0	0	0	1
2	1	1	1	0
3	1	1	1	0

(d)

	0	1	2	3
0	0	0	0	1
1	0	0	0	1
2	0	0	0	1
3	1	1	1	0

(e)

	0	1	2	3
0	0	0	0	1
1	0	0	0	1
2	0	0	0	1
3	0	0	0	1

(f)

Figure 7.5   (a) Memory any in $A(1)$ state; (b) column 3 flipped; (c) row 0 flipped; (d) row 1 flipped; (e) row 2 flipped; (f) row 3 flipped

weight varies from 0 to 3, when the content of the cell is 1. Also, note that cell (3,3) has row weight 3 and column weight from 0 to 3, when the content of the cell is 0.

The next operation is to flip column 2 of Fig. 7.5(f), and then flip all rows one at a time as discussed previously. This is shown in Fig. 7.6. Note that the two corner cells (0,3) and (3,0) have received column weights 0 to 3 with cell value 0 at the end of operation in Fig. 7.6.

Column 1 is flipped next followed by the flipping of all rows as shown in Fig. 7.7. The corner cells (0,0) and (3,3) receive column weights 0 to 3 with cell value 0. Finally, column 0 if flipped; this takes the memory array to $A(1)$ state. At the end of the procedure, each cell with cell value 0 receives varying row and column weights. Note that the row and column weights of the four corner cells vary from 0 to 3. The four corner cells can be tested in a similar manner with cell value 1, by setting the memory array to the $A(0)$ state.

The next step of the algorithm is to test the border cells. This is started by selecting cell $(0, n/2)$. The testing of the cell also results in the simultaneous testing of cells $(0, n/2 - 1)$, $(n - 1, n/2)$, and $(n - 1, n/2 - 1)$. for example, in Fig. 7.8(a) the testing of call (0,2) also results in the testing of cells (0,1), (3,2), and (3,1). This test process partitions the top row into two subrows, the end cells

	0	1	2	3
0	0	0	1	1
1	0	0	1	1
2	0	0	1	1
3	0	0	1	1

(a)

	0	1	2	3
0	1	1	0	0
1	0	0	1	1
2	0	0	1	1
3	0	0	1	1

(b)

	0	1	2	3
0	1	1	0	0
1	1	1	0	0
2	0	0	1	1
3	0	0	1	1

(c)

	0	1	2	3
0	1	1	0	0
1	1	1	0	0
2	1	1	0	0
3	0	0	1	1

(d)

	0	1	2	3
0	1	1	0	0
1	1	1	0	0
2	1	1	0	0
3	1	1	0	0

(e)

Figure 7.6   (a) Column 2 flipped; (b) row 0 flipped; (c) row 1 flipped; (d) row 2 flipped; (e) row 3 flipped

	0	1	2	3
0	1	0	0	0
1	1	0	0	0
2	1	0	0	0
3	1	0	0	0

(a)

	0	1	2	3
0	0	1	1	1
1	1	0	0	0
2	1	0	0	0
3	1	0	0	0

(b)

	0	1	2	3
0	0	1	1	1
1	0	1	1	1
2	1	0	0	0
3	1	0	0	0

(c)

	0	1	2	3
0	0	1	1	1
1	0	1	1	1
2	0	1	1	1
3	1	0	0	0

(d)

	0	1	2	3
0	0	1	1	1
1	0	1	1	1
2	0	1	1	1
3	0	1	1	1

(e)

	0	1	2	3
0	1	1	1	1
1	1	1	1	1
2	1	1	1	1
3	1	1	1	1

(f)

Figure 7.7   (a) Column 1 flipped; (b) row 0 flipped; (c) row 1 flipped; (d) row 2 flipped; (e) row 3 flipped; (f) column 0 flipped

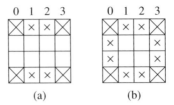

Figure 7.8   Testing of border cells

of each of which have been tested. The subrows can be recursively partitioned in this manner till all the cells in the top row have been tested. The testing of the top row also results in the testing of all cells in the bottom row. Figure 7.8(a) shows the 4 × 4 memory array with the border cells in the top row and the bottom row partitioned into subrows [(0,1), (0,2)] and [(3,1), (3,2)], respectively. The border cells in the leftmost and the rightmost column can be tested similarly by selecting cells (1,0), (2,3), and (1,3). Figure 7.8(b) shows the status of the 4 × 4 memory array after the border cells have been completely tested.

The final step of the testing process is to test the columns $n/2$ and $n/2-1$, and also rows $n/2$ and $n2-1$. This step partitions the memory array into four identical subarrays. Each subarray is similar to the original array with the border cells tested. Figure 7.9(a) shows an 8 × 8 memory array after the corner and the border cells have been tested. The testing of cells in columns 3 and 4, and rows 3 and 4 will result in the partition of the original array into four 2 × 2 subarrays with border cells tested, as shown in Fig. 7.9(b). Each of these subarrays can be partitioned recursively in this manner till all the cells have been tested.

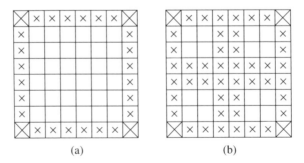

Figure 7.9   Partitioning of the memory array

7.4   BIST Techniques for RAM Chips

Several techniques have been proposed in recent years to incorporate BIST features in RAMs. Dekker *et al.* [7.9] have proposed a scheme to improve the testability of SRAMs by redesigning and augmenting the circuit surrounding the RAM cell array. Figure 7.10 shows the architecture of the testable RAM. The control signals c_1 and c_2 are used to select the mode of operation of the RAM as follows:

c_1	c_2	Mode of Operation
0	0	Normal
0	1	BIST
1	$\times$	Scan

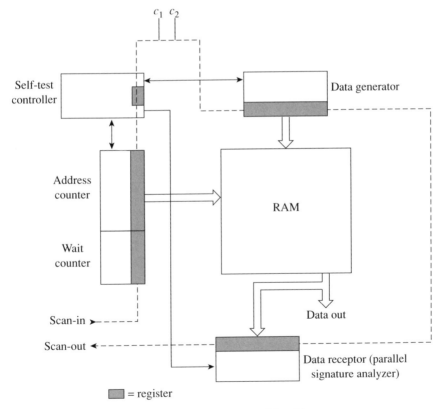

= register

Figure 7.10   Self-testable SRAM architecture

During the BIST mode, the address counter generates the address sequence $0, \ldots, n - 1$ of the RAM array. The wait counter is combined with the address counter, and it is used to test for data retention faults. The detection of such a fault requires that a 0 (1) be written into a memory cell and, after a certain time, the content of the cell be verified. The wait counter generates the necessary amount of wait or delay time for the data retention test.

The function of the data generator is to generate certain data words and their complements in order to detect possible coupling faults between cells at the same address. The data produced by the SRAM during the read operation are accepted by the data receptor (parallel signature analyzer), and compressed to generate a unique signature. Once the signature has been generated, the signature generator is set in a hold mode, and the signature is serially shifted out. It is then compared with the reference signature calculated by using a dedicated software tool. The signature generator consists of a minimum of eight memory elements to ensure a satisfactory fault coverage. The test logic can be fully tested using the built-in scan capability.

Ritter and Muller [7.10] have proposed a BIST scheme in which a built-in processor is used to test and repair large RAMs. The processor consists of a ROM, a RAM, and a processing unit. Figure 7.11 shows the structure of the test processor. The ROM stores 16 constants for testing memory cells, and the RAM is used as a register file during the test program execution. The widths of the words stored in the ROM and the RAM depends on the data word and the address space, respectively, of the memory under test.

The test processor has an instruction set consisting of 11 commands:

LOAD	Read the content memory cells
STORE	Store test pattern
CMP	Compare (used to compare the content of a cell with expected values)
ADD	Add (used to increase cell address)
SUB	Subtract (used to decrease cell address)
MOVE	Register-to-register transfer
NAND	Mask the row and column address
BRA	Branch address
BCF	Branch on carry flag
BNF	Branch on nonzero flag
BZF	Branch on zero flag

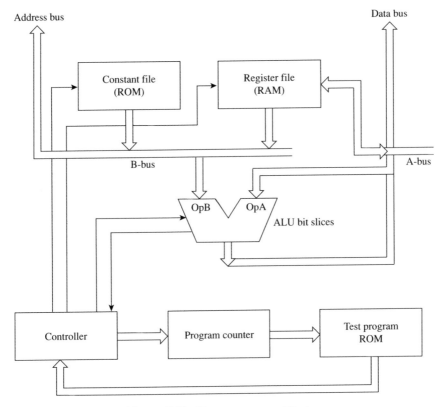

Figure 7.11   Test processor architecture

Each command uses 4 bits of an instruction; the remaining bits of the instruction are determined by the number of registers and constants. The width of the registers and constants are selected based on the organization of the memory under test, that is, address space, data word. Because the lengths of the instructions depend only on the number of registers and constants, not on their widths, the test program is independent of the dimension of the memory under test.

All instructions are carried out in four phases and have equal execution time. For example, the operations involved in reading the content of a cell are as follows:

Phase 1:   The address of the cell is loaded onto the B-bus of the test processor.

Phase 2:   The address is latched into the address decoder; also, the program counter is incremented.

Phase 3:   The next instruction is written into the instruction register of the test processor.

Phase 4:   The contents of the memory location addressed in phase 2 are sent via the data output register onto the A-bus and stored in the register file.

The advantage of the test processor is its flexibility. Any test algorithm can be implemented by the test processor. Also, it can be adapted to different types of memory organization and technology, and it can be used for production tests as well as during system maintenance service.

In general, currently available BIST approaches for memories do not consider repair of memory devices to improve their yield. This method uses up to four spare rows and four spare columns to repair large RAMs. The repair process is based on an intelligent self-test concept. The self-test program is started by initializing the memory cells. Next, test patterns are written into the memory cells. The contents of the cells are then read, and they are compared with expected values. As long as no faulty cell is detected, the test process is continued. If a faulty cell is detected, the corresponding memory address is stored in a register of the test processor. After all fault addresses have been stored, the row and the column with the largest number of faulty cells are replaced. This step is repeated until the spare rows and columns are exhausted. It may not, however, be possible to repair faulty cells, even if spare rows and columns are available, unless the faulty cells are appropriately distributed. If repair of a chip is impossible, the chip is isolated as defective.

The concept of modifying conventional RAM architecture in order to test a number of memory cells simultaneously was introduced by You and Hayes [7.11]. In their technique, a memory array is partitioned into subarrays of size s bits. Each subarray is then reconfigured into an s-bit cyclic shift register. The shift register is recirculated whenever a read operation is performed. The reconfiguration of the subarrays is done by incorporating pass transistors on the bit lines. This, however, results in performance deterioration of a RAM chip.

Sridhar [7.12] proposed a scheme that uses a parallel signature analyzer (PSA) within a RAM to access memory bit lines simultaneously. The PSA can operate in three modes: *scan mode*, *write mode*, and *signature mode*. In the scan mode, the PSA is sequentially loaded with a pattern from outside the memory chip under test. In the write mode, the PSA can write a stored pattern into many bit lines simultaneously. In the signature mode, the PSA reads the contents of memory cells that were previously written into them, and it generates an erroneous sig-

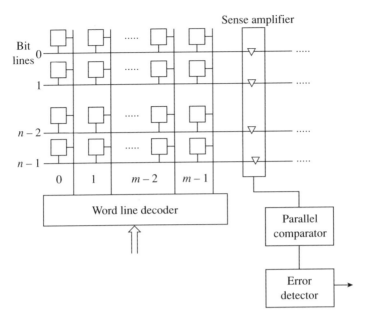

Figure 7.12   Testable DRAM structure

nature if a cell is faulty. However, as in the case of random logic circuits, the
signature analysis approach for testing memory chips introduces the possibility
of aliasing.

Mazumder and Patel [7.13] proposed a technique for designing DRAMs such
that they are amenable to parallel testing of memory cells. Figure 7.12 shows the
structure of a testable DRAM. The memory array has m word lines, each having
n bit lines. The bit lines are partitioned into p blocks such that bit line i belongs
to block b, where $b = i$ mod p. The bit line decoder simultaneously selects all
bit lines in a block during testing. Thus, a WRITE operation in the memory array
while it is in the test mode results in the writing of data in all bit lines in a group
of a selected word line. Similarly, a READ operation during the test mode results
in the reading of data from a group of bit lines in a selected word line. A parallel
comparator is incorporated at the output of the sense amplifiers. The contents of
multiple bit lines accessed by an address are either all 0s or all 1s. However, if a
WRITE operation on a cell fails or the content of some of the simultaneously
accessed cells are faulty, that is, their contents are different from the rest, the
comparator generates an error signal.

7.5   Test Generation and BIST for Embedded RAMs

Embedded RAMs are extremely difficult to test because of poor controllability and observability of their address, control, and data lines. Sachdev and Verstrelen [7.14] have proposed a test algorithm specifically for embedded dynamic RAMs. In this algorithm, the memory array is first initialized to all 0s. The content of each location is then read in ascending order to verify that its content is 0 (assuming bit-oriented RAMs), and a 1 is written into it. This step is repeated to check that the content of each location is 1, and a 0 is written into it. In the next step, each location is read in descending order to check that it contains a 0, and then a 1 is written into it. All memory locations are then read in ascending order to verify that they contain 1s. The memory is disabled for a stipulated time after the reading process is complete, and then read back in ascending order to check data retention for logic 1. A 0 is written into each location after it has been read. The memory is disabled again for a stipulated time, and the locations are read in ascending order to check data retention for logic 0.

Jain and Stroud [7.15] have proposed a BIST technique for testing embedded RAMs. Figure 7.13(a) shows the block diagram of an embedded RAM. For fault modeling purposes a RAM is assumed to be composed of three functional blocks: the memory cell array, the decoder logic, and the read/write logic. The memory cell array is assumed to have one or more of the following faults: single or multiple stuck-at-0/1 faults, coupling faults, data retention faults, and bridging faults between adjacent memory cells. The decoder logic consists of a row address decoder and a column address decoder. A common row decoder but separate column decoders are used for read and write operations. Both single and multiple stuck-at faults in the decoder logic are considered. Multiple stuck-at-faults in a decoder are assumed to make the decoder either access a nonaddressed location, or to access several locations in addition to the addressed location. Because the effect of a multiple fault is similar to a stuck-at fault or a coupling fault in a memory array, it is sufficient to consider the presence of only single stuck-at faults in the decoder logic. Faults in the read/write logic are modeled as stuck-at faults. A stuck-at fault at the output line of a sense amplifier is considered as a stuck-at fault in the memory cell driving the output line. A short or a capacitive coupling between the memory cells and adjacent data input and output lines is considered as a coupling between the memory cells in the columns of the shorted or coupled lines.

The BIST implementation of the embedded RAM of Fig. 7.13(a) is shown in Fig. 7.13(b). A binary counter replaces the address register and the read/write

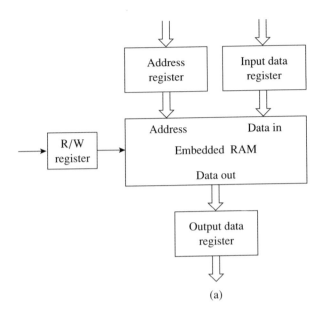

(a)

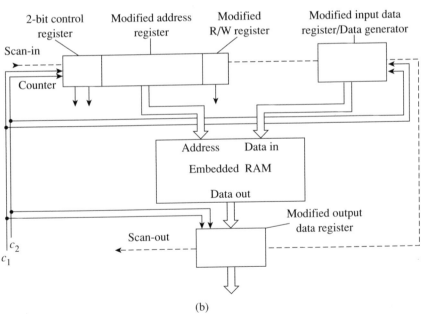

(b)

Figure 7.13   (a) Embedded RAM; (b) equivalent BIST RAM

(R/W) register. The output bits of the counter corresponding to the modified address register are used as address bits for the memory array. The least significant bit of the counter, that is, the output of the modified R/W register, and the two most significant bits of the counter, that is, the output of the control register, are used for controlling the read/write operations of the memory. An additional data generator is included to generate data to be written into the embedded RAM during the self-test mode. The output register is modified to function as a parallel signature analyzer during testing. Also, the counter, the input data register, and the output register can be connected to form a scan path. This path can be used to initialize the counter and the registers, and to test the logic included to incorporate BIST capability into the embedded memory.

The operation mode of the BIST RAM depends on the status of the two test signals c_1 and c_2.

c_1	c_2	Operation Mode
0	0	Normal
0	1	Scan
1	0	Not used
1	1	Self-test

During the self-test mode, the memory is tested by a scheme that stores a checkerboard pattern in the memory array. This test scheme satisfies the necessary conditions for detecting all the assumed faults in the memory array; these conditions are as follows:

Condition 1   A 1 followed by a 0 (or a 0 followed by a 1) is written into each memory cell. The content of a cell must be read after a write operation.

Condition 2   A set of complementary values must be written into two physically adjacent memory cells, and then their complements are written into the cells. For example, if i and j are adjacent memory cells, then a 1 is written into cell i and a 0 into cell j, followed by a 0 in cell i and a 1 in cell j. Each cell must be read after a write operation.

Condition 3   A write operation in a cell (for both 0 and 1) must be followed after a certain delay, known as the *hold time*, by a read operation.

Condition 4   Each memory cell must be read twice after a 1 and a 0 have been written into it.

Condition 5   Every memory location is written into with unique data, and then its content is read to detect faults in the decoders.

Condition 6   A stuck-at fault in the read column decoder logic must be detected using a special test sequence.

Nadeau-Dostie *et al.* [7.16] have proposed a *serial interfacing* technique that enables multiple embedded RAMs inside a chip to share the BIST circuit. This technique is suitable mainly for SRAMs. Figure 7.14 shows the block diagram of an n-word SRAM (with m bits/word). The additional circuit needed to incorporate built-in self-testing is shown in Fig. 7.15. As can be seen in this diagram, the address lines are connected directly to the row decoder and the column decoder. The memory array has separate input and output lines. A word is read from a memory location and transferred to the memory output when the read line is high. The transparent latches at the output of the sense amplifier retains the word when the read line goes back to low. Similarly, when the write line is high, a word is written into a selected RAM location. The BIST circuit has a serial input (SI) and a serial output (SO). The SI is used to shift data into the least significant memory cell of a selected memory location. The data at this input is stored in the BIST circuit till all the memory cells in the addressed location are sequentially loaded with the data. The content of the most significant memory cell is observable at the SO, and it is called the serial output of the memory. The BIST circuit uses only this bit for error detection.

Either the normal data or the test data can be transferred to a memory location depending on the value of the *test mode* (TM) signal. Assuming TM = 0 or 1 corresponds to normal or BIST mode of operation, respectively, let us consider how the content of a memory cell is serially shifted to its adjacent cell during the

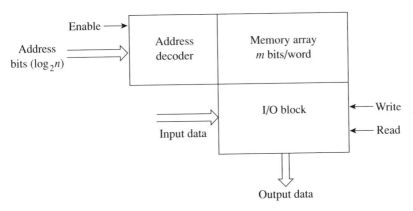

Figure 7.14   Input and output signals for RAM

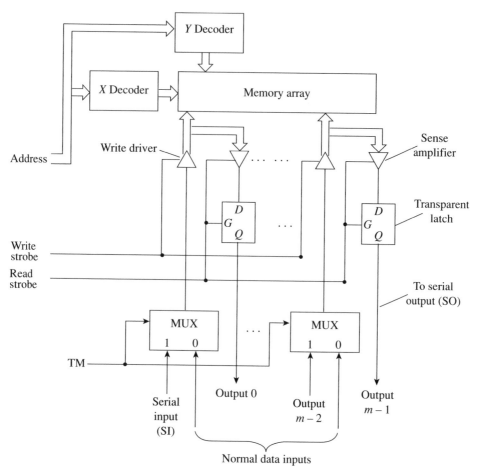

Figure 7.15 Static RAM with serial shifting capability

BIST mode. The address bits of the memory under test are set to select a particular location, and the contents of that location are read. The read line is then set low, which ensures that the data is stored in the latch. Next, the least significant memory cell in the addressed location is written into with the data at the SI. The original content of each memory cells is available at the output of the corresponding latch. A write operation is performed next at the same memory location that was read from, and the content of the least significant cell is stored in the next higher significant cell. This process of successive read and write operations at a particular location is continued till the word at the location is shifted out via SO, and the data at SI replaces all the bits in the location. Most of the memory test algorithms discussed in Sec. 7.2 can be used for testing of embedded RAMs in the serial access method; the serial adaptations of the March, GALPAT, and Walking 0s and 1s algorithms have been provided in Ref. 7.16.

The BIST circuit in the serial BIST scheme can be shared among several RAM blocks inside an ASIC chip by either *daisy chaining* or *test multiplexing*. In daisy chaining, the serial output of one memory block is connected to the serial input of another block as shown in Fig. 7.16. all the RAMs connected in a chain receive the address and control signals simultaneously. The RAMs must have the same number of words if March and GALPAT are used for testing, although the number of bits in a word can be different for different RAMs.

In the test multiplexing method shown in Fig. 7.17, all RAM devices simultaneously receive the address, control, and serial data outputs from the BIST

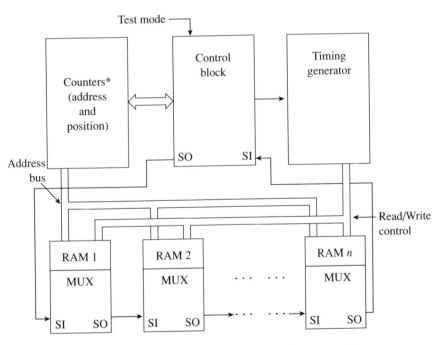

*An address counter has the same number of states as memory addresses, and a position counter has the same number of states as the number of bits in a word.

Figure 7.16   Daisy chaining of embedded RAMs

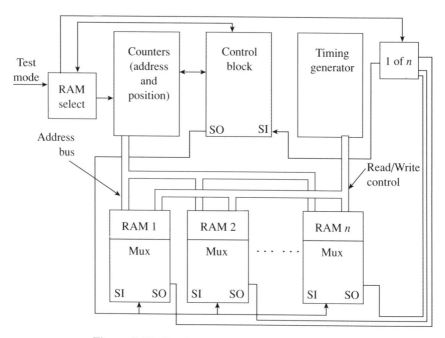

Figure 7.17 Parallel connection of embedded RAMs

circuit. However, one RAM is tested at a time, and its output is applied to the BIST error detection circuit. The untested RAMs remain disabled till they are selected for testing. Unlike in daisy chaining, any of the previously discussed memory test algorithms can be used in this method irrespective of memory dimensions.

Mo *et al.* [7.17] have proposed a BIST structure for embedded SRAMs. Figure 7.18 shows the structure. An up/down counter called the *address counter* is used to access the memory locations in the embedded RAM. The input data to be written into the RAM during the test mode is generated by the *data generator* block. The *data comparator* block compares the data produced by the RAM with expected data during the test mode. If there is an inconsistency in the compared data, an error signal is produced by the comparator. The overall operation of the BIST RAM is controlled by the *self-test controller* block. It detects the status of the address counter, and it prompts the data generator to generate data, and the data comparator to compare data. All the BIST circuit blocks use scan flip-flops that can be connected to form a scan path. The scan path can be used to initialize all the registers, and also to set the status of the self-test controller to apply March

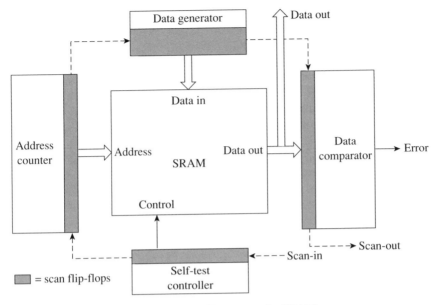

Figure 7.18 BIST structure for SRAMs

and/or checkerboard patterns to test the embedded RAM. Furthermore, the address and the data register contents can be scanned out for the error location if the data comparator detects an error.

References

7.1 Thatte, S., and J. Abraham, "Testing of semiconductor random access memories," *Proc. Intl. Symp. on Fault-Tolerant Computing*, 81–87 (1977).

7.2 Van de Goor, A., "Using March tests to test SRAMs," *IEEE Design and Test of Computers*, 8–14 (March 1993).

7.3 Franklin, M., K. K. Saluja, and K. Kinoshita, "A built-in self-test algorithm for row/column pattern sensitive faults in RAMs," *IEEE Jour. of Solid-State Circuits*, 514–523 (April 1990).

7.4 Sosnowski, J., "In system transparent autodiagnostics of RAMs," *Proc. Intl. Test. Conf.*, 835–844 (1993).

7.5 Breuer, M. A., and A. D. Friedman, *Diagnosis and Reliable Design of Digital Systems*, Computer Science Press (1976).

7.6 Nair, R., "Comments on an optimal algorithm for testing stuck-at faults in random access memories," *IEEE Trans. Comput.*, 258–261 (March 1979).

7.7   Suk, D. S., and S. M. Reddy, "A march test for functional faults in semi-conductor random access memories," *IEEE Trans. Comput.*, 982–985 (December 1981).

7.8   Kinoshita, K., and K. K. Saluja, "Built-in testing of memory using an on-chip compact test scheme," *IEEE Trans. Comput.*, 862–870 (October 1986).

7.9   Dekker, R., F. Beenker, and L. Thijssen, "Realistic built-in self-test for static RAMs," *IEEE Design and Test of Computers*, 26–34 (February 1989).

7.10 Ritter, H. C., and B. Muller, "Built-in test processor for self-testing re-pairable random access memories," *Proc. Intl. Test Conf.*, 1078–1084 (1987).

7.11 You, Y., and J. P. Hayes, "A self-testing dynamic RAM chip," *IEEE Jour. of Solid-State Circuits*, 428–435 (1985).

7.12 Sridhar, T., "New parallel test approach for large memories," *Proc. Intl. Test Conf.*, 462–470 (1985).

7.13 Mazumdar, P., and J. H. Patel, "Parallel testing for pattern sensitive faults in semiconductor random access memories," *IEEE Trans. Comput.*, 394–407 (March 1989).

7.14 Sachdev, M., and M. Verstrelen, "Development of a fault model and test algorithms for embedded DRAMs," *Proc. Intl. Test Conf.*, 815–824 (1993).

7.15 Jain, S., and C. Stroud, "Built-in self-testing of embedded memories," *IEEE Design and Test of Computers*, 27–37 (October 1986).

7.16 Nadeau-Dostie, B., A. Silburt, and V. K. Agarwal, "Serial interfacing for embedded-memory testing," *IEEE Design and Test of Computers*, 57–63 (April 1990).

7.17 Mo, C. T., C. L. Lee, and W. C. Wu, "A self-diagnostic BIST memory design scheme," *Records of Intl. Workshop on Memory Technology, Design and Testing*, 7–9 (August 1994).

Appendix | Markov Models

Markov models are used to analyze probabilistic systems [A.1]. The two key concepts of such models are state and state transition. A system can be in any one of a finite number of states at any instant of time, and move successively from one state to another as time passes. The changes of state are called *state transitions*. In general the probability that a system will be in a particular state at time $t + 1$ depends on the state of the system at time $t, t - 1, t - 2$ and so on. However, if the state of the system at time $t + 1$ only depends on the state at time t, and not on the sequence of transitions through which the state at t was arrived at, the system corresponds to a "first-order" Markov model. If the state at time $t + 1$ is independent of the previous state, that is, the state at time t, then the system corresponds to a Markov model of "zero order."

The probability that the system will, when in state i, make a transition to state j is known as the *transition probability*. A system with S states has S^2 transition probabilities, which can be denoted by $p_{ij}, 1 \leq i \leq S, 1 \leq j \leq S$. For computational purposes the transition probabilities are organized into a square matrix P, called the *transition probability matrix*, as shown below; the (i, j) entry in P is the probability of transition from state i to state j.

$$P = \begin{bmatrix} p_{11} & p_{12} & \cdots & p_{1S} \\ p_{21} & p_{22} & \cdots & p_{2S} \\ \vdots & \vdots & & \vdots \\ p_{S1} & p_{S2} & \cdots & p_{SS} \end{bmatrix}$$

In any particular situation, the transition probabilities p_{ij}, and consequently the transition matrix P, depend upon what is assumed about the behavior of the system.

The Markov models discussed so far are "discrete-time" models. These models require all state transitions to occur at fixed intervals: each transition is assigned with a certain probability. On the other hand "continuous-time" Markov models are characterized by the fact that state transitions can occur at any point in time; the amount of time spent in each state, before proceeding to the next state, is exponentially distributed.

Reference

A.1 Shooman, M. L., *Probabilistic Reliability: an Engineering Approach*, McGraw-Hill (1968).

Index

A

Address counter, 190
Addressable latch, 122
Algebraic factoring, 62
Algebraic product, 61
Aliasing, 157
Appearance fault, 75
ASIC, 169
ATE, 140

B

Backtrace, 44
Backward trace, 22
Base cells, 174
BILBO, 162
Bit line, 170
Boolean difference, 23
Boolean product, 61
Boundary scan cells, 133
Bounding, 42
Built-in self-test (BIST), 140
Bypass register, 136

C

Cellular automation, 151
Checkerboard test, 174

Checking experiments, 81
Checking sequence, 81
Checkpoints, 52
Circular BIST, 159
Circular flip-flops, 159
Classical fault model, 2
Complement-free ring sum expansion, 56
Constant weight counter, 144
Controllability, 101, 132, 133
Controllable flip-flops, 129
Convolved LFSR/SR, 147
Coupled cell, 171
Coupling fault, 170
Coverage probability, 16
CrossCheck, 130
CrossCheck grid, 131
Cross-controlled latch, 132
Cross-point faults, 74
Cube, 61
Cube-free expression, 64

D

Daisy chaining, 189
D-algorithm, 29
Data comparator, 190
Data generator, 190
Data retention faults, 172
D-Drive, 34

195